Applications of Genetics to Arthropods of Biological Control Significance

Edited by

S. Karl Narang
Alan C. Bartlett
Robert M. Faust

CRC Press
Taylor & Francis Group
Boca Raton London New York

CRC Press is an imprint of the
Taylor & Francis Group, an **informa** business

First published 1993 by CRC Press
Taylor & Francis Group
6000 Broken Sound Parkway NW, Suite 300
Boca Raton, FL 33487-2742

Reissued 2018 by CRC Press

Library of Congress Cataloging-in-Publication Data

Applications of genetics to arthropods of biological control significance / editors,
S. Karl Narang, Alan C. Bartlett, Robert M. Faust.
p. cm.
Papers presented at a workshop held at the 19th International Congress of Entomology.
Beijing, China. 1992.
Includes bibliographical references and index.
ISBN 0-8493-2607-9
1. Arthropods as biological pest control agents—Congresses. 2. Arthropods—Genetics—Congresses. I. Narang, S. Karl. II. Bartlett, Alan. C. III. Faust, Robert, M.
IV. International Congress of Entomology (19th: Peking, China: 1992)
SB976.A78A66 1994
632'.7—dc20 93-44973

A Library of Congress record exists under LC control number: 93044973

ISBN 13: 978-1-315-89066-1 (hbk)
ISBN 13: 978-1-351-06976-2 (ebk)

Visit the Taylor & Francis Web site at http://www.taylorandfrancis.com and the
CRC Press Web site at http://www.crcpress.com

Big-eyed bug, *Geocoris punctipes* (Say). Photo courtesy of Jack Dykinga.

EDITORS

S. Karl Narang, Ph.D. is a Research Leader of the Insect Genetics and Cellular Biology Research Unit at the U.S. Department of Agriculture, Agricultural Research Service, Fargo, ND. Dr. Narang received his genetics training at the University of Illinois, Urbana-Champaign, receiving the M.S. degree in 1968 and the Ph.D. degree in 1970. He served as a Senior Scientist/Geneticist at the World Health Organization's Project on Genetic Control (New Delhi) in 1971; as Associate Professor and later Chairman of the Department of Animal Biology, University of Brasilia, Brasilia, Brazil from 1972 to 1979; Research Scientist at the Department of Entomology, University of Florida, Gainesville from 1980 to 1985; and Research Geneticist at the USDA, ARS laboratory in Gainesville from 1986 to 1989. It was in 1990 that he assumed his present position.

Dr. Narang is a member of the Genetic Society of America, American Genetics Society (life member), Entomological Society of America, Florida Entomological Society (serving as subject-editor in genetics), American Mosquito Control Association, Panjab University Zoological Society (life member), has been a member of the American Association of Advancement of Science, New York Academy of Science, Sigma Xi, Brazilian Genetics Society, Brazilian Society for the Advancement of Science, and Brazilian Biomedical Association. He has served as a consultant and technical advisor to the National Academy of Science, National Research Council, Board of Science and Technology for International Development, and reviewed overseas projects on genetics and ecology of insect pests. He has served as consultant to the U.S.-Brazil Biomedical Research Collaboration, University of Maryland and University of Bahia, Salvador, Bahia, Brazil. He has received several awards including Gold medals for highest academic achievements during his B.S. and M.Sc. studies, Fullbright Travel award, World Health Organization Fellowship award, and *WHO's WHO in Frontiers of Science and Technology* (1985, 2nd edition of Marquis *WHO's WHO*). He has been the recipient of many research grants for the National Research Council and Walter Reed Army Institute of Research. He has organized many workshops and chaired symposia sponsored by FAO/IAEA/DOE/USDA. the Rockefeller Foundation, and by many national and international societies.

Dr. Narang is the author of more than 120 papers and has co-edited 2 books and contributed 8 book chapters. His current major research interests relate to population genetic structures of insect pests and genetic engineering of insects as they relate to implementation of biological control programs.

Alan C. Bartlett, Ph.D., is a Research Geneticist (insects) at the U.S. Department of Agriculture, Agricultural Research Service, Western Cotton Research Laboratory, Phoenix, AZ. Dr. Bartlett received an Associate in Science degree from Carbon College (now College of Eastern Utah), Price, UT (1954), his B.A. (cum laude) and M.S. degrees in genetics at the University of Utah, Salt Lake City, UT (1956, 1957) and his Ph.D. in population and radiation genetics at Purdue University, Lafayette, IN (1962). He was an instructor at Carbon College in 1957–1958 and at Purdue University in 1960–1962. He joined USDA/ARS as an insect geneticist in 1962 at the

Boll Weevil Research Laboratory, State College (now Mississippi State University), MS. In 1967 he transferred to the Cotton Insects Research Laboratory in Tucson, AZ and in 1970 he assumed his present position as Research Geneticist in Phoenix, AZ. He served as an adjunct professor of genetics at Mississippi State University and Arizona State University, Tempe, AZ.

Dr. Bartlett is a member of the Genetics Society of America, Entomological Society of America, Southwestern Entomological Society, Sigma Xi, and American Institute of Biological Sciences, and has been a member of AAAS, American Genetic Association, Phi Sigma, and the Arizona Academy of Science. He is listed in Marquis *Who's Who in America*, *Who's Who in Science and Technology*, and *Who's Who in the West*. He has organized and participated in symposia sponsored by the Entomological Society of America, International Atomic Energy Agency, Hungarian Academy of Science, International Crops Research Institute for the Semi-Arid Tropics, USDA, ARS, the Arizona Academy of Sciences, and the International Congress of Entomology . He served as president of the Mississippi State University Chapter of Sigma Xi and chairman of the Tempe Gifted Association. He received the Outstanding Alumnus award from the College of Eastern Utah in February 1993.

Dr. Bartlett is the author of more than 80 papers, the co-editor of 1 book, and has contributedchapters to 8 books. His current major research interests are the development of genetic methods of control of insects that attack cotton, the development of genetic methods of identification of geographic or host races of insects, and the genetic effects of the domestication process on insects.

Robert M. Faust, Ph.D., is a National Program Leader of fundamental and molecular biology on the National Program Staff of the U.S. Department of Agriculture, Agricultural Research Service, Beltsville, MD. Dr. Faust received his training in microbiology and entomology at the University of Maryland, College Park, receiving his B.S. in 1963, his M.S. in 1966, and Ph.D. in 1970. He served at the USDA/ARS, Market Quality Research Laboratory in 1959–1960, the Bioenvironmental Bee Laboratory, 1960–1988, and has been in his present position since 1988.

Dr. Faust is a founding member of the Society of Invertebrate Pathology, a member of the American Society of Microbiology, the Entomological Society of America, AAAS, New York Academy of Science, Japanese Society of Sericultural Science, Biochemical Society of the Republic of Korea, Korean Society of Applied Microbiology, American Registry of Professional Entomologists, American Legion, and Sigma Xi. Dr. Faust was a visiting scientist at the Armenian Academy of Science Institute of Microbiology, USSR, 1976; Korean Institute of Science and Technology, Seoul, 1985; part-time faculty of Prince Georges Community College, Largo, MD, 1964–1984; and president, SCH Corporation, 1977. He served on the executive board of the Center of Biological Control of Mosquitoes, Ben Gurion University, 1981. Dr. Faust has served on the editorial board of the Journal of Invertebrate Pathology (1980–1985) and authored books and articles on such varied subjects as *Diseases of Insects*, 1974; *Microbial and Viral Insecticides*, 1982; *Genetic Engineering in the Plant Sciences*, 1981; *The Molecular Biology of the Bacilli*, 1985; and contributed critical reviews and scientific articles to professional journals.

PREFACE

The continued use of broad-spectrum synthetic chemical insecticides over more than four decades has led to widespread insect resistance, as well as heightened the concern over their adverse effect on the environment and human health. Further, the U.S. Environmental Protection Agency has banned the use of many such chemicals and severely restricted the use of others; more are likely to be banned. Consequently, the discovery and development of environmentally sound, publicly acceptable, and viable economical strategies of pest control have taken on a new urgency. An important foundation of pest control is integrated pest management (IPM), of which an important component includes the use of biological control agents; IPM offers one of the best long-term strategies to help reduce or eliminate the use of synthetic chemical pesticides.

Biological control is based on the use of selected natural enemies from the native range of the pest to reduce the pest's negative effects. Generally, it includes (1) importation and release of natural enemies, (2) augmentation of populations of established biotypes by laboratory mass rearing, and (3) conservation of natural enemies by manipulation of the habitats. However, biological control technology is considered by many to be still in its infancy. The literature contains many examples of both successes and failures. The failure of biological programs to reduce the intended effect can be traced to an inadequate knowledge and understanding of fundamental pest life processes, biosystematics, genetics, ecological influences; and pest, natural enemies, host interrelationships. Thus, there are many questions that must be resolved before biological control strategies can reach a higher level of success as an integral component of IPM-based systems approaches to pest control.

First, it is critical that we know the specific status of each target pest and the natural enemies of that pest. For example, many parasitic species have not yet been identified and named, and many of these are sibling species. The presence of unrecognized species in a collection might account for the partial (or even lack of) success of some biocontrol programs. Some investigators question the need for concern about genome diversity in biocontrol organisms. The identification of cryptic species is important in two respects. Different species may attack different hosts. If they do attack the same host, they may differ in their attributes for biological control. There are many examples where failure to detect cryptic taxa in the target pest or its biocontrol agents has led to failure of biocontrol projects. It should be evident to the reader that detecting cryptic taxa is not a trivial matter, it can mean the difference between success and failure.

Populations of the same species may differ in relevant biological characteristics important to biological control. Therefore, biological and genetic variation within and among species must be characterized to develop effective biocontrol strategies. Such studies become even more critical when new exotic enemies are imported and their establishment in the new environment forms the basis of the program.

Many areas and crops have been or are currently being treated with pesticides that can destroy the parasitoid as effectively as it does the pest species. One way to solve this problem would be to use parasitoids that are resistant to the pesticide. This would allow the release of parasitoids in combination with pesticide application, thus increasing the effectiveness of both. Extensive work is being done in a few laboratories to select for resistant strains from field populations. This is an expensive and laborious task and has been criticized for this reason. Development of pesticide-resistant strains is one avenue that may be exploited using biotechnological techniques.

Augmentation of natural enemies involves mass production and periodic colonization under laboratory conditions. When a colony is established from a sample of a natural population collected abroad, only some of the genetic variability of the area of origin is represented. This may account for the partial success of some biocontrol agents, but some investigators question the need to be concerned about genetic diversity. Many, if not most, of the highly successful biological control projects have occurred from limited introduction that would appear to represent only a fraction of species genome diversity. This would suggest that other factors, such as exclusion of pathogens, might take priority over concern about genome diversity as exotic natural enemies are chosen and prepared for field release. The successful establishment and spread of the vedalia beetle, *Rodolia cardinalis*, in California from one introduction is the example often cited in support of this view. That success has been repeated often following introduction of the vedalia beetle from California into other countries. Often, a small subset of the original California introduction was included in the subsequent releases.

Contrary to the success of the vedalia beetle in some parts of California is the failure or only partial success of this predator in coastal California and Bermuda. Was genome diversity too limited for the insect to perform adequately under cooler climatic conditions? Are there biotypes which would perform better under cooler climates? There are no easy answers to such questions. Adequate data is lacking to assess the breadth of genomic diversity necessary to ensure effectiveness of natural enemies. Therefore, development of methods to determine the role of genome diversity (and to increase it, if necessary) in the effectiveness of natural enemies should be done to ensure success of biological control strategies.

Recent advances in biotechnology provide tremendous potential for improvement of the effectiveness of natural enemies against insect pests. This could range from the development of genetic sexing strains to engineering of disease and pesticide resistance for beneficial insects. However, genes controlling desirable traits must be identified. Next, these genes must be cloned and characterized. Finally, these genes must be suitably modified and successfully introduced into the target organism's DNA such that the desired trait will be transmitted in a Mendelian fashion.

It was against this background that a symposium/workshop on "Genetic Applications to Arthropods of Biological Control Significance" was held at the XIX International Congress of Entomology in Beijing, China in the summer of 1992. This symposium/workshop brought together scientists from federal and state research institutions, both foreign and domestic, to review the state of knowledge, to identify research gaps and needs, and to suggest corresponding approaches to meet the needs of this important approach to the use of parasitoids and predators in biological control. It was recognized that there is a paucity of information on the genetics of natural enemies in the literature. Therefore, this book compiles papers presented during the symposium/workshop and also articles by other contributors who were unable to attend. We believe this book details the current state of knowledge and makes an earnest attempt at addressing many of the basic issues in terms of genetic innovations that have the potential to considerably improve the effectiveness of modern biological control programs.

S. K. Narang
A. C. Bartlett
R. M. Faust

CONTRIBUTORS

Alan C. Bartlett
Western Cotton Research Laboratory
U.S. Department of Agriculture
Agricultural Research Service
Phoenix, AZ

Fred D. Bennett (Retired)
Entomology and Nematology
Department
University of Florida
Gainesville, FL
and
Crofton, Baldhoon Road
Laxey, Isle of Man, Great Britain

Allen C. Cohen
Western Cotton Research Laboratory
U.S. Department of Agriculture
Agricultural Research Service
Phoenix, AZ

J. D. DeVault
Biosciences Research Laboratory
U.S. Department of Agriculture
Agricultural Research Service
University Station
Fargo, ND

R. M. Faust
National Program Staff
U.S. Department of Agriculture
Building 005
BARC-West
Beltsville, MD

Dan Gerling
Department of Zoology
The George S. Wise Faculty
of Life Sciences
Tel Aviv University
Ramat Aviv, Israel

L. J. Heilmann
Biosciences Research Laboratory
U.S. Department of Agriculture
Agricultural Research Service
University Station
Fargo, ND

Marshall W. Johnson
Department of Entomology
University of Hawaii at Manoa
Honolulu, HI

C. M. Krueger
Department of Medical Biochemistry
and Genetics
Health Science Center
College of Medicine
Texas A & M University
440 Reynolds Medical Building
College Station, TX

R. A. Leopold
Biosciences Research Laboratory
U.S. Department of Agriculture
Agricultural Research Service
University Station
Fargo, ND

N. P. Markwick
The Horticulture and Food Research
Institute of New Zealand, Ltd.
Auckland, New Zealand

S. K. Narang
Biosciences Research Laboratory
U.S. Department of Agriculture
Agricultural Research Service
University Station
Fargo, ND

Robin E. Owen
Department of Chemical and
Biological Sciences
Mount Royal College
Calgary, Alberta, Canada
and
Department of Biological Sciences
University of Calgary
Calgary, Alberta, Canada

Robert T. Staten
Methods Development
U.S. Department of Agriculture
APHIS
Phoenix, AZ

William W. M. Steiner
Biological Control of Insects
Research Laboratory
U.S. Department of Agriculture
Agricultural Research Service
Columbia, MO

W. J. Tabachnick
Arthropod-Borne Animal Diseases
Research Laboratory
U.S. Department of Agriculture
Agricultural Research Service
Laramie, WY

Bruce E. Tabashnik
Department of Entomology
University of Hawaii at Manoa
Honolulu, HI

Gennaro Viggiani
Dipartimento di Entomologia
e Zoologia Agraria,
Università di Napoli Federico II
Via Università,
Portici (NA), Italy

TABLE OF CONTENTS

Chapter 1

GENETICS AND INSECT BIOTYPES: EVOLUTIONARY AND PRACTICAL IMPLICATIONS*

William W. M. Steiner

TABLE OF CONTENTS

* All programs and services of the U.S. Department of Agriculture are offered on a nondiscriminatory basis without regard to race, color, national origin, religion, sex, age, marital status, or handicap. Mention of any proprietary products in this chapter does not signify endorsement by the U.S. Department of Agriculture.

0-8493-2607-9/94/$0.00+$.50

"How should the taxonomist treat clones, pure lines, biotypes, and so-called 'strains' or 'stocks' of parthenogenetic or asexual organisms?"
Methods and Principles of Systematic Zoology[1]

I. INTRODUCTION

In their now classic book, Mayr, Linsley, and Usinger[1] discussed the roots of modern taxonomy. They point out, in particular, how the abandonment of the typological species concept which derived at least from the time of Linnaeus was weakened by the acceptance of evolutionary thought and then replaced by a more dynamic, polytypic concept. Here, emphasis was placed on the study of variation within populations and quantitative analysis of the differences themselves.[1] Shortly after the turn of the century, species were "...no longer considered as something fixed and uniform, but rather as polytypic, consisting of many subspecies and local populations, each differing from the others and each showing considerable variability within itself... ."[1] It became clearly recognized that the differences among subspecies were made up of many small variations. Studies of these differences can be seen to continue today in the form of isozyme and molecular analyses. In addition, it became clear that the environment played a role; variation in local populations of a species could be correlated with differences in local environments. Studies of this realization continue today as well, in the interests displayed by population geneticists in clines in gene frequencies.

Nevertheless, questions did arise. How were biologists to classify the different variants seen in natural populations, and of what significance might these be? Taxonomically, the species has been considered as the lowest common denominator for a group of individuals with the potential to reproduce with each other. However, the variants themselves deserved some form of recognition, since they could be used to describe the limits and distribution of a local population or group of populations. Various names did arise, including subspecies, race, microspecies, strain, variant, clone, line, and biotype. In insects, the latter term gained some significance.

More than a decade ago, Gonzalez et al.[2] succinctly summarized the problem that insect biotypes presented for biological control of pest insects. A key point was that the bases for discriminating between biotypes were not firm, primarily due to differences among taxonomists regarding definitions of different taxonomic levels and because of a serious lack of genetically based information by which to discriminate different taxa. These authors also pointed out that the discrimination of biotypes was especially important for insect pest management programs involving resistance management and host attraction. Berlocher[3] pointed out that although electrophoresis and the techniques to establish biochemically distinct taxonomic categories had begun to prove useful, the conceptual basis for identifying subspecific taxa (host races, subspecies, biotypes) was not well developed due to a lack of discriminatory studies at these levels as well as the problem of interpretation which Gonzalez et al.[2] had identified.

The problem of undefinable taxa goes much deeper than considerations of potential roles in pest management. For example, Ryder[4] points out the importance of subspecies to species and gene pool conservation. The problem is very important with respect to the role zoos now play in maintaining diversity in endangered animal species as seen in the five extant tiger subspecies or the seven extant black rhino subspecies. In this case, zookeepers must identify which subspecies "...actually represent populations possessing genetic attributes significant for present and future generations... ." The same can be said for entomologists interested in maintaining the genetic diversity of beneficial insects. In this case, not only must we identify gene pools of significant interest, we must understand those biotic and abiotic factors that make it important in the agroecosystem.

In this chapter, the latter two problems are discussed in some detail. Where others have speculated on the importance of biotypes in the context of biological control, I would like to examine them from an evolutionary viewpoint. In so doing, I will try to show not only what the nature of the biotype is, but the implications this holds for genetic engineering programs in particular, and insect control in general.

II. DEFINITIONS: SUPERSPECIFIC TO BIOTYPIC RANKS

When taxonomists freed biology of the monotypic species concept, it became necessary to distinguish the broad range of variation which could be encountered in any one species. Not so apparent were the relationships that a variable species might have to the varying components of the environment experienced by different populations across its range and the genetic complexities which might underlie the variation observed to exist in the species itself.

Nevertheless, sense needed to be made of this variation which is now recognized to underlie the "polytypic" species concept and has some importance in helping to define the boundaries of what constitutes the biotype. Using the superspecies as the starting point, Mayr, Linsley, and Usinger[1] gave us the following definitions which are more or less standard and are useful here.

A. SUPERSPECIES

Superspecies are "...a monophyletic group of very closely related and largely or entirely allopatric species."[1] This definition recognizes that reproductively isolated species exist and that they form a closely related, monophyletic group easily distinguishable from other groups either morphologically or genetically. This classification has little bearing on the concept of biotypes, since the latter may not fall into either morphologically or genetically recognizable classes. However, it may be that, in the evolution of new species (formation of the species complex), biotype formation may play an as yet undefined role, possibly through continual recolonization of viable habitats as described by Frey, Latscha, and McKey[5] or in sympatric speciation models such as proposed by Muller.[6] In Frey, Latscha, and McKey's[5] model, the concepts of Carson (References 7 and 8 and papers cited therein) may play a role if the host undergoing colonization is viewed as an "island" of colonization opportunity. In this scenario, several biotypes of a single species form the basis of the monophyletic grouping if enough genetic changes accrue in the manner outlined by Carson.[7,8] Evidence of such an occurrence might be forthcoming from ecological and genetic studies done simultaneously on monophyletic aphid species, even though Carson's models are derived and discussed specifically for sexually reproducing species.

B. SPECIES

Species are "...groups of actually or potentially interbreeding natural populations which are reproductively isolated from other such groups."[1] This definition is based on the biological criterion of reproductive isolation. Not withstanding the challenges which have been made to the very idea of species in a classification scheme,[9,10] there are two basic types of species to be recognized: the monotypic species, which are species with low levels of subspeciation and population differentiation, and the polytypic species, which are species consisting of two or more subspecies. Gene flow is seen as a major conservative force in the former. In taxa containing subspecies, it is assumed that gene flow is restricted by geographical barriers of some kind and that monotypic species lack such restrictions, although Jackson and Pounds[11] commented that critical studies remain to be done to prove the apparent truism. Assumptions concerning gene flow do not recognize inherent differences in internal genetic architecture underlying genome organization which could alter the ability of a population to differentiate. For example, Jackson and Pounds[11] pointed out that chromosomal differentiation may present a barrier to gene flow.

Templeton[12] has identified some of the salient effects of the genetic architecture underlying gene flow in his summary of the recent concepts associated with the genetics of speciation. He recognizes at least two major modes: speciation due to divergence and speciation due to transilience. Templeton defines these as "...(in) divergence, the isolating barriers evolve in a continuous ... fashion, with some form of natural selection, either directly or indirectly, being the driving force leading to reproductive isolation. Transilient modes involve a discontinuity in which some sort of selective barrier is overcome by other evolutionary forces."[12]

Templeton[12] goes on to define the mechanisms for three separate types of models (adaptive, clinal, and habitat based) for the divergence mode and four types of models (genetic, chromosomal, hybrid maintenance, and hybrid recombination) for the transilience mode. These models are important for our consideration because biotypes might arise in one or more models as a *prerequisite* to formal speciation, but as a part of the microevolutionary process. Examination of the conditions outlined by Templeton for each model to occur suggests that divergence models might be more important to biotype formation than transilience models and that genetic transilience may be the most important of the transilience models due to the nature and consequence of pleiotropic interactions at the genetic level. This latter model would necessarily assume that a host plant shift in a biotype, for example, constitutes a founding event accompanied by associated shifts in the genes underlying coadaptation, but preceding speciation.

C. SIBLING SPECIES

Sibling species are "...pairs or groups of very similar and closely related species ... (which) do not differ from other species in any respect except for the minuteness of their structural differences."[1] We now know that such species groups may not necessarily be closely related genetically, but the difficulty in telling the species in the group apart is nevertheless still great. Often, the criteria for differentiation is based on reproductive isolation leading to definable genetic differences as described for the malaria mosquito *Anopheles nuneztovari*.[13] The use of reproductive isolation described by Narang and Seawright[14] for identifying sibling species of mosquitoes is applicable for any species and has long been used in identifying sibling species of *Drosophila*. The consideration of sibling species is important here because Muller[6] has found prezygotic and postzygotic reproductive isolation mechanisms among plant host races of aphid species, suggesting that some biotypes may in actuality be sibling species.

D. SUBSPECIES

Subspecies are "...geographically defined aggregates of local populations which differ taxonomically from other such subdivisions of a species."[1] Any discussion about gene flow is relevant here, since differentiation is difficult when it is present, and likely when it is absent. Although it was the opinion of Gonzalez et al.[2] that the term subspecies remain undefined, it is the topic of extensive discussion in Mayr, Linsley, and Usinger.[1] As Gonzalez et al. point out, extensive controversy has existed over the meaning and definition of this and other infrasubspecific names such as ecophenotypes, ecotypes, forms, host races, strains, stocks, and varieties for some time. Although some definitions exist for these various categories (e.g., Berlocher[3] restricts the term strain to "...insect populations manipulated by man"; Bush[15] defined host race as "...(populations) which show preference for a host which is different from the host or hosts of other populations of the same species"; Gregor,[16] following Turesson's 1922 treatise, defined the ecotype as "...the product arising as a result of the genotypical response of an ecospecies to a particular habitat"; and Mayr, Linsley, and Usinger[1] defined "geographical race" as synonomous with "subspecies"), where reproductive isolation is not present. All infrasubspecific taxa actually fall into the category of subspecies.

It is necessary, however, to point out that biotype and subspecies should not be used interchangeably. Mayr, Linsley, and Usinger[1] are precise in their definition of subspecies and point out that this is a defined taxonomic category under the International Code of Zoological Nomenclature.[17] No assumption is made of a genotype/environment interaction for a subspecies, but this interaction is necessary for identification of the biotype. In subspecies, genetic differentiation may be a result of isolation and reduced gene flow coupled with genetic drift as well as selection. Thus, subspecies should be reserved for the broader interpretation of within-species differentiation.

E. BIOTYPES

Whereas Mayr, Linsley, and Usinger[1] seem to classify biotypes in the class of clones, lines, and "strains" of parthenogenetic or asexual organisms, Muller[6] accepts the definition of Von Keler:[18] "Biotype, biotypus, pure type, pure line are all genotypically equal isogenous specimens of a population procreated by autogamy or

TABLE 1
The Determination Grid Attempts to Identify and Separate Different Taxa Based on Reproductive and Genetic Characteristics and Demonstrates the Appropriate Placement of Biotypes

Morphology	Not reproductively isolated	Genetically differentiated	Reproductively isolated	Genetically differentiated
Identical				
Sympatric	(1) Same population; no host differences	No	(5) Sibling sp.	Yes
	(1b) Same population; host differences	Yes (biotype)		
Allopatric	(2) Same subspecies; no host differences	No	(6) Sibling sp.	Yes
	(2b) Different subspecies; host differences	Yes (biotype)		
Different				
Sympatric	(3) Individual variants of same population; no host differences	No	(7) Different sp.	Yes
	(3b) Individual variants of same population; host differences	Yes (biotype)		
Allopatric	(4) Different subspecies; no host differences, but other differences	No	(8) Different sp.	Yes
	(4b) Different subspecies; host differences	Yes (biotype)		

Note: It is assumed that niche or host specificity is known in all cases.

Modified after Mayr, E., Linsley, E. G., and Usinger, R. L., *Methods and Principles of Systematic Zoology*, McGraw-Hill, New York, 1953.

parthenogenesis. The biotype can rarely be regarded as a systematic category. Genetically, biotypes are slight mutants and therefore represent hereditary variations. Environmental factors may select ecotypes from among the biotypes of a species. Biotype richness in a species is a sign of the youthfulness and vigour of the species, of its plasticity for adaption to new biotopes, a sign that the species is flourishing." Mayr, Linsley, and Usinger[1] then proposed a way to identify biotypes based on the comparison of some simple genetic and reproductive characteristics (Table 1).

Note that Mayr, Linsley, and Usinger[1] and Von Keler[18] both associate the biotype with autogamous or parthenogenetic (self-fertilizing or asexual) reproduction. However, aphids are rarely strictly parthenogenic; more commonly they are thelytokus. Also, "biotype" has since escaped the narrower definition and now includes sexually reproducing strains of species as well (e.g., Costa et al.[19] for sweetpotato whitefly; Shukle, Grover, and Mocelin[20] for Hessian fly). I propose here the following definition which takes the wider interpretation into account. *Biotypes are genetic variants of a species which share a common genetic base predisposing their carriers to a higher genetic fitness within a specific and identifiable ecological niche.* The differences in hereditary variation remain at the root of the biotype in both definitions, but no assumptions are made about reproductive mode or plasticity for change. This is seen in the two examples offered above, where esterase differences (*Bemisia tabaci* Gennadius) and virulent vs. avirulent genotypes (*Mayetiola destructor* Say) parallel the type of genetically based differences seen in aphid biotypes (see Mettler[21] for a brief review of genetically based nutritional differences). The wider interpretation suggests a way to classify particular genetic variants of a species, leaving the broader implications of genetic selection open to critical assessment. Other sexually reproducing species that show complete or partial genetic differences which might be host dependent can be found in Table 2. Host dependency offers an environmental difference which serves as a starting point for the dissection of the basis of selection in biotypes, a point recognized by Nemec and Stary.[22]

The acceptance of these criteria (genetic and niche-specific differences regardless of morphology) enables one to argue that even parasitoids, where they demonstrate host-specific differences, may have biotypes (see Table 2, and Nemec and Stary[23] for such a case in the aphid parasitoid *Diaeretiella rapae* M'Int., where esterase differences were found between strains reared from different aphid hosts). It is clear, if we accept that an ecologically dependent genetic difference underlies the biotypes seen within a species, that implications now exist for genotype/environment interactions and evolution. Although Muller[6] suggests that sympatric speciation may be involved in evolution of biotypes, this is not the only interpretation. We now turn to a consideration of these implications.

III. GENETIC ORIGINS OF THE BIOTYPE

The genetic origins of the biotype may be varied and are based on selection theory. Typically, biotypes have been associated with nutritional differences in aphids for some time,[21,24] and this probably gave some impetus to research connecting host plant differences with aphid biotypes. It is clear, however, that nutritional factors may not be all that is involved. For example, fitness differences have been shown for fecundity.[25-28] In addition, Puterka and Peters[29] have studied the heritability and genetics of biotype virulence in the aphid herbivore *Schizaphis graminum* (Rondani). In this species, a duplicate gene-modifier gene inheritance was postulated with avirulence dominant to virulence.

Genetics probably lies at the base of differences in virulence or virus vector capacity,[25,30,31] photoperiod and lifecycle reactions,[32] structural biochemical differences,[33] temperature susceptibility,[34] susceptibility to resistance genes in plants,[35-41] and adaptation.[42]

In some cases, definite structural genetic differences have been identified at the levels of the chromosome,[43-45] nuclear and mitochondrial DNA,[46-48] and ribosomal DNA.[49] With the exception of the study by Beranek and Berry[50] who studied *Aphis fabae* (Scopoli) and *Myzus persicae* (Sulz.), the genetic basis of the polymorphic

TABLE 2
Sexually Producing Species Which Show Intraspecific Host Dependency Plus Genetic Differentiation, Thus Meeting the Criteria for Assignation of Biotype Status

Species	Genetic differences	Host specificity	Ref.
Plant pests			
Anthonomus grandis	mtDNA fragments	*Gossypium hirsutum* *G. thurberi* *Hampea nutricia*	74
Bemisia tabaci	β esterases	Varied, primary concern is differences in vector capability	75
Biston betularia [a]	Melanic genes	Varied, depends on background coloration	76
Mayetiola destructor	Virulence interaction	*Triticum aestivum*	20
Orseolia oryzae	Plant resistance genes	*Oryza sativa*	38
Phytomyza chaerophylli	Pgm allele frequencies	*Chaerophyllum temulum* L.	5
Rhagoletis pomonella	*Acon-2* and *Had* allele frequencies	*Malus pumila* *Crataegus mollis*	66
Schizaphis graminum	*Gb2* and *Gb3* plant resistance genes	Various cereals and grasses	29
Diuraphis noxia	RAPD-PCR, PGI, EST, PGDH	Various cereals and grasses	69
Entomophagous parasites			
Diaeretiella rapae	Isoesterase frequencies	*Brevicoryne brassicae* *Hayhurstia atriplicis* *Myzus persicae*	23
Comperiella bifasciata	Superparasitism and ability to escape encapsulation	*Aonidiella aurantii*	77
Pediobius foveolatus	*Acon-1* and *Mdh-1*	*Epilachna* spp.	78

[a] The morphological trait of interest is industrial melanism, occurring in many moth species of which this is but one example; according to our definition of biotype, a clear ecological/genetic interaction underlies these cases.

enzymes observed has not been generally established in aphids due to their primarily parthenogenetic mode of reproduction. Nevertheless, isozymes have been shown to be related to host plant and other differences.[51,52] It should be pointed out that isozymes may show no or reduced variation in parthenogenetic forms[53,54] and so may have limited applicability with respect to serving as markers to differentiate biotypes. Still, a model of how biotypes arise from the genome is necessary. Here, the robust, multigenic model proposed by Puterka and Peters[29,30] has a significant impact because it implies that hundreds of biotypes may be possible by genetic recombination. They assume that biotype predominance is determined by selection pressure such as host availability. This would seem to fit the observations made by many workers in the field.

Although morphological characters have been used to distinguish biotypes, the genetic basis underlying these is not always clear or substantiated. Thus, Kolesova, Kuznetsova, and Shaposhnikov[55] examined an array of morphological markers differentiating Russian clones of *M. persicae* reared on tobacco, potatoes, and carrots. Although longer length of different appendages and rostral segments could be used to differentiate carrot biotypes, other body parts demonstrated no significant differences. Partitions of the variance were not attempted. Although the relative contributions of genetics vs. environmental variability is not clear in the latter study, it is evident that quantitative traits involving many genes can play a role. These types of differences have not been studied in great detail where biotypes have been recognized.

Discontinuous variation rightly falls in the category of "polymorphism" as defined by Clark.[56] Clark identified two types of polymorphism where the fact of variation is known but its causation is not: "genetically determined polymorphism" and "environmentally cued polymorphism". In the former, the genotype is paramount in determining the morph; in the latter, a distinct genotype/environment interaction exists with the genotype, resulting in response to an environmental cue or trigger. Supposedly, environmentally cued polymorphism "produces morphs congruent with current ecological conditions". If this type of polymorphism is present, it might mislead identification of a biotype, since Clark[56] assumes that an all-responsive genome is present and environmental differences bring out morph differences. There is, in other words, no selection present to tailor a genome to a particular niche; rather, homeostatic buffering mechanisms resulting in different morphologies are the root of niche adaptation. It would appear, then, that genetically determined polymorphisms coupled with selection pressure underlies the origin of biotype differences and that the number of genes involved can vary.

IV. ECOLOGICAL IMPORTANCE OF THE BIOTYPE

Muller[57] has succinctly pointed out the role of the aphid biotype (he used the term subspecies, but the cases he describes clearly involve host dependency, genetics, and niche specificity) in applied entomology. In several species complexes, he gives examples where one type within a species may be a vector of disease, but another may not because of inherently different biological susceptibilities, type of disease vectored, or behavior including host specificity that precludes exposure to a particular plant virus. These differences extend to color (morphology) and to reproduction as well, providing a way to test biological susceptibility on the different hosts.[57] Indeed, Steiner[58] has summarized the evidence suggesting how differences in susceptibility to infection might act to promote niche and genetic differentiation and thus promote evolution. Thus, from the agricultural point of view, it becomes important to be able to identify the biotype with respect to host specificity and ability to vector disease.

To the biological control advocate, however, another reason becomes immediately apparent. It is feasible that aphid parasitoids may attack a range of biotypes within a species, unless host specificity becomes predominant as is found in *D. rapae*.[23] In the case of a broad host range, nonvirulent biotypes of a host aphid may serve to maintain the beneficial insect at economically important densities. This may be true for predatory insects such as coccinellids as well. The role nonvirulent biotypes may play in maintaining beneficial insects has never been investigated formally, and in the shift of emphasis from insecticidal control to integrated pest management (IPM) and biological control of arthropod pests seen in todays agroenvironment, it is a role worth investigating. It may be that one of the best ways to augment the natural occurrence of an entomophagous parasite or predator is to simply seed harmless biotypes of aphids into the desired environment in order to obtain higher densities of a biocontrol agent.

It must be emphasized that since the identification of the biotype is closely linked with genotype/environment interaction, recognition and characterization of the ecological niche is of extreme importance. The majority of cases reported in the literature tend to report only a host-specific relationship. In particular, information on biochemical and physiological interactions are missing or they have not been studied. It is precisely these interactions which might yield clues as to how the adaptive process leading to biotype formation resulted in the genotype/environment interaction which is observed. Powers et al.[59] have described models in which genetic mechanisms may underly adaptation to a different or changing environment; these are worth consideration in order to understand how biotypes might be formed.

Biotype discrimination and its importance in biological control has already been discussed by Gonzalez et al.[2] Of major consideration are the problems inherent in identifying the correct biotype of an entomophagous insect for biological control purposes. Obviously, it is important to establish the ecological context in which both the pest insect and the desired biological control agent occur and interact. Matching these will be of the utmost importance to the success of a biological control program.

V. ROLE OF THE BIOTYPE IN EVOLUTION

Muller's[6] suggestion that "biotypes … may be good for the first step of evolutionary divergence…" is worth serious consideration. For the applied entomologist, this suggestion has serious implications rooted in coevolutionary theory. For example, it suggests that introduced parasitoid species should be matched with their specific host biotype whenever possible to ensure success, simply because the parasite has developed mechanisms to circumvent host defenses. However, the relationships of importance may be much more subtle than first indicated. For example, Sokal and Riska,[60] in a multivariate analysis of two morphs of *Pemphigus populitransversus*, found that the environmental factors to which the morphs are adapting appear to have different autocorrelation patterns, suggesting that different microevolutionary processes are involved. The nature of these processes remains undiscovered.

Still, some insights can be gained by studying the evolution of the host/parasitoid relationship. Stary, Lyon, and Leclant[61] summarize studies which demonstrate that the development of host specificity very often is accompanied by a lowering of fitness if the parasitoid is switched to a new or even previous host and that the adaptive process might occur very rapidly and be enhanced where the parasitoid is an introduced exotic species. Such an expansion characterized by new host adaptations might occur, even though the founding population may have suffered some loss of genetic diversity via founding events. Such a finding might be construed as providing support for the evolutionary models offered by Carson[7,8] who suggests that passage of a population through a bottleneck may promote a major shift in genetic balance. Such a shift is envisioned as providing new epistatic genetic variance, which "makes novel evolution possible" under new selective pressures. Indeed, the occurrence of specialization in diet may be much more common than at first believed, belying the widely held idea that diet breadth is a species characteristic (Fox and Morrow;[62] see also the interesting arguments developed by Jackson and Pounds[11] which may have a bearing here). If this is true, then literally hundreds of species identified to have biotypes actually may be demonstrating incipient speciation *in situ*, the biotype serving as the transition phase so long sought after in studies of evolution. Evidence supporting this position might be forthcoming if the following appropriate questions are asked. Do monophyletic or monotypic groups of organisms demonstrate fewer instances of biotypic differentiation than polyphyletic or polytypic groups? Do sibling species complexes demonstrate more biotypic differentiation than species groups not having sibling species or having them present in low numbers? The extent to which niche specialization plays a role in the answer to these questions would also be important.

The differentiation of gene pools of either pest or beneficial insects is central to the interpretation of evolutionary events. However, such interpretation must be made carefully, for they carry serious consequences whether wrong or right. Thus, in at least one case, biotypic status was alluded, which was not warranted and which could have led to costly mistakes in biological control decisions.[63,64] In another, sibling species rather than biotypic status is implicated, indicating that recognition of the appropriate species could signal whether or not control is required.[65] The same might be true of the biotypes known as "host races" in *Rhagoletis pomonella*, the apple maggot, if timing of emergence is enough to reproductively isolate the gene pools of adults flies.[66] These examples and others are enough to point out that evolutionary considerations can affect insect management programs, an impact which cannot be taken lightly.

The host-specific differentiation observed in some "parasites" (Werren et al.[67] for the effect of gypsy moth host plant on *Cotesia melanoscela*; Gilbert and Coaker[68] for considering *Pieris rapae* as a parasite of different brassica cultivars; Singh and Painter[34] for *Rhopalosiphum maidis* on different host plants) offer unique opportunities to study adaptation in process. Little is understood about which physiologic and metabolic systems might be involved, how they are affected, and with what they interact within the host. Yet, at no other time in biological sciences have better techniques been available to investigate these problems. Insight into these problems could clearly yield an advantage in terms of insect control.

VI. IMPLICATIONS OF THE BIOTYPE IN PEST MANAGEMENT AND GENETIC ENGINEERING PROSPECTS

There are at least two ways in which the biotype influences insect management programs. One involves the immediate approach or treatment to be considered if agroeconomics is seriously impacted.[2] As pointed out earlier, a prime consideration here is to simply match the right pest and beneficial biotypes to each other; this is considered by Gonzalez et al.[2] to be primarily a problem of taxonomy with respect to infraspecificity. This problem extends into host-plant preferences as well. Here, better ways to discriminate among host-specific biotypes are needed, and the new polymerase chain reaction (PCR) techniques which discriminate small DNA differences between individual insects can play a key role.[46,69] PCR techniques are sensitive enough that DNA probes can be designed around key markers to enable quick field identification of biotypes of interest.

The second way is the consideration of how to develop future methods and approaches or even prevent future crises from arising. Understanding the interaction of host-plant molecules with pest or beneficial biochemical systems, for example, might enable unique genetic engineering approaches to be considered in which special molecules are constructed for specific attractions or interactions. Already, unique programs exist for design and modification of specific protein molecules in food,[70] and with the right knowledge concerning what molecules cause host specificity to occur, one might consider programs which modify the plant host just enough to make it unrecognizable to the pest biotype. Such programs would be independent of those envisioned by Gould[71] for engineering of pest resistance genes in plants, or agricultural geneticists might engineer the attractant into a trap crop meant to fool the pest into attacking it rather than a food plant. These approaches, in the long run, may be much more compatible than the current idea of insertion of naturally occurring insecticidal genes such as those found in the bacteria *Bacillus thuringiensis* (*Bt*) into plants simply because there would then be no danger of loss of the natural insecticides ability to kill due to intensified selection pressure favoring resistant insect biotypes.

Alternatively, one might design plant genes to be better at attracting beneficial insects to them when damaged. For example, parasitoids of the genus *Microplitis* (Hymenoptera: Braconidae) attacking *Heliothis/Helicoverpa* moth species are known to orient to plant kairomones issuing from pest-damaged plants.[72,73] Enhancing levels of these compounds in the appropriate host plant might ensure a quicker and surer response from appropriate beneficial parasitic biotypes.

VII. CONCLUSIONS AND SUMMARY

Here, the definition of the biotype from its original restriction to parthenogenetic forms is expanded to include sexually reproducing forms. The expansion seems justified on the grounds that the connotation of the word biotype will always imply a broader semantic meaning, as demonstrated in the way various researchers have begun to use it. Since even parthenogenetic forms occasionally go through a sexual phase, the contribution of this phase to biotype formation cannot be estimated, but is probably substantial, as it would provide the recombination field necessary to generate the new genotypes upon which niche-specific selection might operate. The interested reader is reminded that subspecies are a distinct and recognized taxonomic category, and it is pointed out that biotypes may be members of subspecific groups, but that subspecies should not be considered biotypes. Recognition of biotypic status can be made by applying the concepts as modified from Mayr, Linsley, and Usinger[1] and presented in Table 1.

Although the genetic origins of the biotype may be varied and monogenic or polygenic depending on the nature of the genotype/environment interaction tying the interacting genomes together, fitness differences will ultimately arise in response to niche-specific selection pressure. These differences will enhance within-niche selection and might result in determinable genetic differences represented in the literature as isozyme, DNA, or chromosomal differences between biotypes. The determination of these differences is not always guaranteed; however, and it is suggested that the new PCR techniques and probe development will better enable biotype discrimination. Such discrimination is important to biological control approaches in order to be sure that the right host biotype has been matched with the right pest control agent biotype.

It is pointed out that identification and characterization of the ecological niche is of extreme importance, not only in recognizing a biotype which might occupy it, but also in order to understand the basis of the specificity which enables the insect to recognize its niche. This specificity itself can be put to good use if we understand it; I suggest that natural population augmentation of parasitoids with a broad host range which includes virulent hosts might be done by seeding nonvirulent (harmless) host biotypes into a local environment as a food and development source for the parasitoid. In addition, since the genotype/environment interaction underlies its development, our understanding of biotype evolution can be enhanced by understanding the ecological niche in its proper perspective, that of selective agent.

It must be understood that coevolution of the host (plant or insect herbivore) and the parasitoid (insect) really underlies the evolution of the biotype. The differentiation of the respective gene pools which are involved in the interaction can be studied with great detail today. The characterization of this differentiation offers unique opportunities to study adaptation in process. Understanding the nature of physiological and biochemical changes which are or have taken place can yield insight into how beneficial insects interact with their niche (plant or insect host) and might give us protein or kairomone systems to target for further development. This development might take one of several forms. The applied geneticist/entomologist might modify the plant host just enough to make it kairomonally more attractive to a beneficial insect or unrecognizable to a pest (herbivore) biotype. On the other hand, perhaps simultaneously with the former technique, the gene(s) responsible for the attraction chemical might be moved to a weedy trap crop, thus fooling the pest to go elsewhere for its dinner. These approaches should be taken as serious alternatives to engineering natural insecticides into crop plants, especially where the effective loss of the insecticide is indicated by the development of insecticide resistance in the targeted pest organism. It goes without saying that the success of these alternative approaches might very well rely on the strength of the genotype/environment interaction underlying biotypic specificity.

ACKNOWLEDGMENTS

I would like to thank Drs. Robert Sites and Gary Puterka who offered valuable comments and criticisms of the original manuscript and Drs. Alan Bartlett and Sudhir K. Narang for the encouragement to put these perspectives on paper. The insights offered here, however, remain my own.

REFERENCES

1. **Mayr, E., Linsley, E. G., and Usinger, R. L.,** *Methods and Principles of Systematic Zoology*, McGraw-Hill, New York, 1953.
2. **Gonzalez, D., Gordh, G., Thompson, S. N., and Adler, J.,** Biotype discrimination and its importance to biological control, in *Genetics in Relation to Insect Management*, Hoy, M. A. and McKelvey, J. J., Jr., Eds., Rockefeller Foundation Publications, New York, 1979, 129.
3. **Berlocher, S. J.,** Biochemical approaches to strain, race, and species discriminations, in *Genetics in Relation to Insect Management*, Hoy, M. A. and McKelvey, J. J., Jr., Eds., Rockefeller Foundation Publications, New York, 1979, 137.
4. **Ryder, O. A.,** Species conservation and systematics: the dilemma of subspecies, *TREE*, 1, 9, 1986.
5. **Frey, M. E., Latscha, T., and McKey, D. B.,** Genetic differentiation and speciation in leaf-mining flies of the genus *Phytomyza*, *Entomol. Exp. Appl.*, 57, 191, 1990.
6. **Muller, F. P.,** Biotype formation and sympatric speciation in aphids (Homoptera: Aphidinea), *Entomol. Gener.*, 10, 161, 1985.
7. **Carson, H. L.,** Genetic variability and the founder effect on remote oceanic islands, in *Biogeographical Aspects of Insularity*, Proc. Int. Symp., Roma Accademia Nazionale Dei Lincei, Atti Dei Convegni Lincei, Rome, 85, 741, 1990.
8. **Carson, H. L.,** Genetic change after colonization, *GeoJournal*, 28, 297, 1992.
9. **Lovtrup, S.,** The evolutionary species concept. Fact or fiction?, *Syst. Zool.*, 28, 386, 1979.
10. **Wiley, E. O.,** Is the evolutionary species fiction? — A consideration of classes, individuals and historical entities, *Syst. Zool.*, 29, 76, 1980.
11. **Jackson, J. F. and Pounds, J. A.,** Comments on assessing the dedifferentiating effect of gene flow, *Syst. Zool.*, 28, 78, 1979.
12. **Templeton, A. R.,** Mechanisms of speciation — a population genetic approach, *Annu. Rev. Ecol. Syst.*, 12, 23, 1981.
13. **Steiner, W. W. M., Kitzmiller, J. B., and Osterbur, D. L.,** Gene differences in chromosome races of *Anopheles nuneztovari* (Gabaldon), *Mosq. Syst.*, 12, 306, 1980.
14. **Narang, S. K. and Seawright, J. A.,** Electrophoretic method for recognition of sibling species of Anopheline mosquitoes a practical approach, *Fla. Entomol.*, 71, 303, 1988.
15. **Bush, G. L.,** Sympatric host race formation and speciation in frugiverous fruit-flies of the genus *Rhagoletis* (Diptera: Tephritidae), *Evolution*, 23, 237, 1969.
16. **Gregor, J. W.,** The ecotype, *Biol. Rev.*, 19, 20, 1944.
17. **Richter, R.,** *Einfuhrung in die Zoologische Nomenkltur durch Erlauterung der Internationalen Regeln*, W. Kramer, Frankfurt, 1948.

18. **Von Keller, St.,** *Entomologisches Worterbuch.* Vol. 3, Auflage, Akademie-Verlag, Berlin, Deutsch Democratisches Republic, 1963.
19. **Costa, H. S., Johnson, M. W., Ullman, D. E., Omer, A. D., and Tabashnik, B. E.,** Sweetpotato whitefly (Homoptera: Aleyrodidae): analysis of biotypes and distribution in Hawaii, *Environ. Entomol.,* 22, 16, 1993.
20. **Shukle, R. H., Grover, P. B., Jr., and Mocelin, G.,** Responses of susceptible and resistant wheat associated with Hessian fly (Diptera: Cecidomyiidae) infestations, *Environ. Entomol.,* 21, 845, 1992.
21. **Mettler, T. E.,** Application of artificial feeding techniques for aphids, in *Aphids, Their Biology, Natural Enemies and Control,* Vol. B, Minks, A. K. and Harrewijn, P., Eds., Elsevier, Amsterdam, 1988, 145.
22. **Nemec, V. and Stary, P.,** Genetic diversity and host alternation in aphid parasitoids (Hymenoptera: Aphidiidae), *Entomol. Gener.,* 10, 253, 1985.
23. **Nemec, V. and Stary, P.,** Population diversity of *Diaeretiella rapae* (M'Int.) (Hym., Aphidiidae), an aphid parasitoid in agroecosystems, *A Angew. Entomol.,* 97, 223, 1984.
24. **Srivastava, P. N. and Auclair, J. L.,** Differential responses of biotypes of the pea aphid, *Acyrthosipon pisum* (Harris), to a chemically defined diet, *Can. J. Zool.,* 56, 2481, 1978.
25. **Wilhoit, L. R. and Mittler, T. E.,** Biotypes and clonal variation in greenbug (Homoptera: Aphididae) populations from a locality in California, *Environ. Entomol.,* 20, 757, 1991.
26. **Service, P.,** Genotypic interactions in an aphid-host plant relationship: *Uroleucon rudbeckiae* and *Rudbeckia laciniata, Oecologia,* 61, 271, 1984.
27. **Markkula, M. and Roukka, K.,** Resistance of plants to the pea aphid, *Acyrthosiphon pisum* Harris. I. Fecundity of the biotypes on different host plants, *Ann. Agric. Fenn.,* 9, 127, 1970.
28. **Painter, R. H. and Pathak, M. D.,** The distinguishing features and significance of the four biotypes of the corn leaf aphid *Rhopalosiphum maidis* (Fitch), *Proc. Int. Congr. Entomol.,* 11, 110, 1962.
29. **Puterka, G. J. and Peters, D. C.,** Inheritance of greenbug, *Schizaphis graminum* (Rondani), virulence to *Gb2* and *Gb3* resistance genes in wheat, *Genome,* 32, 109, 1989.
30. **Puterka, G. J. and Peters, D. C.,** Sexual reproduction and inheritance of virulence in the Greenbug, *Schizaphis graminum* (Rondani), in *Aphid-Plant Genotype Interactions,* Campbell, R. K., and Eikenberry, R. D., Eds., Elsevier, New York, 1990, 289.
31. **Thottappilly, G., Eastop, V. F., and Bath, J. E.,** Morphological variation within *Acyrthosipon pisum* and its ability to transmit broadbean severe chlorosis virus, *Entomol. Exp. Appl.,* 22, 29, 1977.
32. **Chmyr, P. G. and Kolesova, D. A.,** Photoperiodic reactions of the pea aphid *Acyrthosipon pisum* (Homoptera, Aphididae) from different food plants, *Zool. Zh.,* 68, 466, 1989.
33. **Dillwith, J. W., Bergman, D. K., Fargo, W. S., Puterka, G. J., and Peters D. C.,** Cuticular hydrocarbons of greenbug biotypes, in *Proc. of Aphid-Plant Interactions: Populations to Molecules Conference,* Peters, D. C., Webster, J. A., and Chlouber, C. S., Eds., Oklahoma State University and USDA Joint Publication, No. MP-132 , Stillwater, OK, 1990, 297.
34. **Singh, S. R. and Painter, R. H.,** Effect of temperature and host plants on progeny production of four biotypes of corn leaf aphid, *Rhopalosiphum maidis, J. Econ. Entomol.,* 57, 348, 1964.
35. **McCauley, G. W., Jr., Margolies, D. C., and Reese, J. C.,** Field asessment of greenbug (Homoptera: Aphididae) demography on corn, *Environ. Entomol.,* 21, 1072, 1992.
36. **Dixon, A. G. O., Bramel-Cox, P. J., and Harvey, T. J.,** Diallel analysis of resistance in sorghum to greenbug biotype E: antibiosis and tolerance, *Crop Sci.,* 30, 1055, 1990.
37. **Puterka, G. J. and Burton, R. L.,** Aphid genetics in relation to host plant resistance, in *Proc. of Aphid-Plant Interactions: Populations to Molecules Conference,* Peters, D. C., Webster, J. A., and Chlouber, C. S., Eds., Oklahoma State University and USDA Joint Publication, No. MP-132, Stillwater, OK, 1990, 59.
38. **Rao, P. S. P. and Kittur, S. U.,** Characterization of Cuttack and Raipur biotypes of the Asian rice gall midge in India, *Trop. Pest Manag.,* 35, 428, 1989.
39. **Kindler, S. D., Staples, R., Spomer, S. M., and Adeniji, O.,** Resistance of bluegrass cultivars to biotypes C and E greenbug (Homoptera: Aphididae), *J. Econ. Entomol.,* 76, 1103, 1983.

40. **Nielson, M. W. and Kuehl, R. O.,** Screening efficacy of spotted alfalfa aphid biotypes and genic systems for resistance in alfalfa, *Environ. Entomol.*, 11, 989, 1982.
41. **Cartier, J. J. and Painter, R. H.,** Differential reactions of two biotypes of the corn leaf aphid to resistant and susceptible varieties, hybrids and selections of sorghums, *J. Econ. Entomol.*, 49, 498, 1956.
42. **Weber, G.,** On the ecological genetics of *Sitobion avenae* (F.) (Hemiptera, Aphididae), *Z. Angew. Entomol.*, 100, 100, 1985.
43. **Mayo, Z. B., Starks, K. J., Bands, D. J., and Veal, R. A.,** Variation in chromosome length among five biotypes of the greenbug (Homoptera: Aphididae), *Ann. Entomol. Soc. Am.*, 81, 128, 1988.
44. **Mayo, Z. B. and Starks, K. J.,** Chromosome comparisons of biotypes of *Schizaphis graminum* to one another and to *Rhopaloziphum maidis, R. padi.* and *Sipha flava*, *Ann. Entomol. Soc. Am.*, 65, 925, 1972.
45. **Blackman, R. L.,** Chromosomes and parthenogenesis in aphids, in *Insect Cytogenetics*, Blackman, R. L., Hewitt, G. M., and Ashburner, M., Eds., Blackwell Scientific Publications, Oxford, 1980, 133.
46. **Black, W. C., IV,** Use of molecular probes to characterize aphid populations, in *Proc. of Aphid-Plant Interactions: Populations to Molecules Conference*, Peters, D. C., Webster, J. A., and Chlouber, C. S., Eds., Oklahoma State University and USDA Joint Publication, No. MP-132, Stillwater, OK, 1990, 137.
47. **Carvalho, G. A., Wratten, S. D., and MacLean, N.,** Differentiation of aphid clones using DNA fingerprints obtained from individual aphids, in *Aphid-Plant Interactions: Populations to Molecules*, Peters, D. C., Webster, J. A., and Chlouber, C. S., Eds., Proc. of a Conference, Oklahoma State University and USDA Joint Publication, No. MP-132, Stillwater, OK, 1990, 287.
48. **Powell, T. O., Jensen, S. G., Kindler, S. D., Stryker, C. J., and Sandall, L. G.,** Mitochondrial DNA divergence among greenbug (Homoptera: Aphididae) biotypes, *Ann. Entomol. Soc. Am.*, 82, 298, 1989.
49. **Lupoli, R., Irwin, M. E., and Vossbrinck, C. R.,** A ribosomal DNA probe to distinguish populations of *Rhopalosiphum maidis* (Homoptera: Aphididae), *Ann. Appl. Biol.*, 117, 3, 1990.
50. **Beranek, A. P. and Berry, R. J.,** Inherited changes in enzyme patterns within parthenogenetic clones of *Aphis fabae* and *Myzus persicae*, *Entomol. Exp. Appl.*, 17, 129, 1974.
51. **Abid, H. S., Kindler, S. D., Jensen, S. G., Thomas-Compton, M. A., and Spomer, S. M.,** Isozyme characterization of sorghum aphid species and greenbug biotypes (Homoptera: Aphididae), *Ann. Entomol. Soc. Am.*, 82, 303, 1989.
52. **Steiner, W. W. M., Voegtlin, D. J., Irwin, M. E., and Kampmeier, G.,** Electrophoretic comparison of aphid species: detecting differences based on taxonomic status and host plant, *Comp. Biochem. Physiol. B*, 81, 295, 1985.
53. **Suomalainen, E., Saura, A., Lokki, J., and Teeri, T.,** Genetic polymorphism and evolution in parthenogenetic animals. IX. Absence of variation within parthenogenetic aphid clones, *Theor. Appl. Genet.*, 57, 129, 1980.
54. **May, B. and Holbrook, F. R.,** Absence of genetic variability in the green peach aphid, *Myzus persicae* (Hemiptera: Aphididae), *Ann. Entomol. Soc. Am.*, 71, 809, 1978.
55. **Kolesova, D. A., Kuznetsova, V. G., and Shaposhnikov, G. Kh.,** Clonal variability in *Myzus persicae* Sulz. (Homoptera, Aphididae), *Entomol. Rev.*, 59, 21, 1980.
56. **Clark, W. C.,** The environment and the genotype in polymorphism, *Zool. J. Linn. Soc.*, 58, 255, 1976.
57. **Muller, F. P.,** The role of subspecies in aphids for affairs of applied entomology, *J. Appl. Entomol.*, 101, 295, 1986.
58. **Steiner, W. W. M.,** Parasitization and speciation in mosquitoes: a hypothesis, In *Cytogenetics and Genetics of Vectors*, Pal, R., Kitzmiller, J. G., and Kanda, T., Eds., Proc. Symp. XVI Int. Congr. Entomol., Kyoto, New York, 1981, 91.
59. **Powers, D. A., Lauerman, T., Crawford, D., and DiMichele, L.,** Genetic mechanisms for adapting to a changing environment, *Annu. Rev. Genet.*, 25, 629, 1991.
60. **Sokal, R. R. and Riska, B.,** Geographic variation in *Pemphigus populitransversus* (Insecta: Aphididae), *Biol. J. Linn. Soc.*, 15, 201, 1981.

61. **Stary, P., Lyon, J. P., and Leclant, F.,** Post-colonisation host range of *Lysiphlebus testaceipes* in the Mediterranean area (Hymenoptera, Aphidiidae), *Acta Entomol. Bohemoslav.*, 85, 1, 1988.
62. **Fox, L. R. and Morrow, P. A.,** Specialization: species property or local phenomenon?, *Science*, 211, 887, 1981.
63. **LaChance, L. E., Bartlett, A. C., Bram, R. A., Gagne, R. J., Graham, O. H., McInis, D. O., Whitten, C. J., and Seawright, J. A.,** Mating types in screwworm populations?, *Science*, 218, 1142, 1982.
64. **Richardson, R. H., Ellison, J. R., and Averhoff, W. W.,** Reply to LaChance *et al.*, *Science*, 218, 1143, 1982.
65. **Pashley, D. P.,** Host-associated genetic differentiation in fall armyworm (Lepidoptera: Noctuidae): a sibling species complex?, *Ann. Entomol. Soc. Am.*, 79, 898, 1986.
66. **McPheron, B. A., Courtney Smith, D., and Berlocher, S. H.,** Genetic differences between host races of *Rhagoletis pomonella, Nature (London)*, 336, 64, 1988.
67. **Werren, J. H., Raupp, M. J., Sadoff, C. S., and Odell, T. M.,** Host plant used by gypsy moths affect survival and development of the parasitoid *Cotesia melanoscela, Environ. Entomol.*, 21, 173, 1992.
68. **Gilbert, N. and Coaker, T. H.,** Differential survival of British and Mediteranean strains of *pieris rapae* (L.) (Lepidoptera: Pieridae) on different brassica cultivars, *Bull. Entomol. Res.*, 78, 669, 1988.
69. **Puterka, G. J., Black, W. C., IV, Steiner, W. M., and Burton, R. L.,** Genetic variation and phylogenetic relationships among worldwide collections of the Russian wheat aphid, *Diuraphis noxia* (Mordvilko), inferred from allozyme and RAPD-PCR markers, *Heredity*, 70, 604,1993.
70. **Creamer, L. K., Jimeniz-Flores, R., and Richardson, T.,** Genetic modification of food proteins, *Trends Biotech.*, 6, 163, 1988.
71. **Gould, F.,** Evolutionary biology and genetically engineered crops, *Bioscience*, 38, 26, 1988.
72. **Herard, F., Keller, M. A., Lewis, W. J., and Tumlinson, J. H.,** Beneficial arthropod behavior mediated by airborne semiochemicals. IV. Influence of host diet on host-oriented flight chamber responses of *Microplitis demoliter* Wilkinson, *J. Chem. Ecol.*, 14, 1597, 1988.
73. **Lewis, W. J. and Jones, R. L.,** Substance that stimulates host-seeking by *Microplitis croceipes* (Hym.: Braconidae), a parasite of *Heliothis* species, *Ann. Entomol. Soc. Am.*, 64, 471, 1971.
74. **Roehrdanz, R. L. and North, D. T.,** Mitochondrial DNA restriction fragment variation and biosystematics of the boll weevil, Anthonomus grandis, *Southwest. Entomol.*, 17, 101, 1992.
75. **Costa, H. S. and Brown, J. S.,** Variation in biological characteristics and esterase patterns among populations of *Bemisia tabaci*, and the association of one population with silverleaf symptom induction, *Entomol. Exp. Appl.*, 61, 211, 1991.
76. **Kettlewell, H. B. D.,** Phenomenon of industrial melanism in lepidoptera, *Annu. Rev. Entomol.*, 6, 245, 1961.
77. **Blumberg, D. and Luck, R. F.,** Differences in the rates of superparasitism between two strains of *Comperiella bifasciata* (Howard) (Hymenoptera: Encyrtidae) parasitizing California red scale (Homoptera: Diaspididae): an adaptation to circumvent encapsulation?, *Ann. Entomol. Soc. Am.*, 83, 591, 1990.
78. **Hung, A. C. F. and Schaeffer, P. W.,** Isozyme analysis of six populations of *Pediobius foveolatus* (Crawford) (Hymenoptera: Eulophidae), *Proc. Entomol. Soc. Wash.*, 92, 160, 1990.

Chapter 2

COMPLEXITIES OF POPULATION GENETIC STRUCTURE AND IMPLICATIONS FOR BIOLOGICAL CONTROL PROGRAMS

S. K. Narang, W. J. Tabachnick, and R. M. Faust

TABLE OF CONTENTS

0-8493-2607-9/94/$0.00+$.50

I. INTRODUCTION

During the past decade, research on traditional biological control (parasites, predators, and microbes) and the application of this technology to pest management has expanded markedly because this strategy of control has been increasingly attractive as an alternative to the use of synthetic organic chemical pesticides. DeBach and Rosen[1] have provided a comprehensive account of relevant biological and ecological principles, and they have recommended research approaches aimed at improving biological control strategies.

Traditional biological control involves three fundamental approaches: (1) discovery, importation, evaluation, and release of exotic natural enemies; (2) augmentation of populations of established biotypes by laboratory mass colonization; and (3) conservation of natural enemies. To enhance the effectiveness of these approaches, it is critical to ensure that there is a single species and not a species complex involved in the release. It is important for the natural enemy species and the target pest to be identified. There are many examples where failure to detect cryptic taxa of the target pest or the biocontrol agents being used has led to a failure of the effort. According to Paterson,[2] "Detecting cryptic taxa is not just an academic luxury, it is often the difference between success and failure, effectiveness and bumbling."

Although much has been published in terms of the importance of correct species identification and the dangers of incorrect identification on the success of biological control, it is important to realize the danger of too much emphasis on a typological view of the species concept. The species concept is important when species differ in important biological traits critical to the biological control effort. Intraspecies variation, however, is important, and populations of the same species may differ in relevant biological characteristics of importance to biological control. It is variation, within and among species, which must be characterized. To ensure success of a biological control program, adequate knowledge of (1) biosystematics; (2) biogeography; (3) biology, i.e., behavior, developmental cycles, physiology, reproduction, nutrition, culture, and genetics; and (4) ecology and population biology of the pest and its natural enemies is critical. Genetic improvement of natural enemies to enhance desirable traits will be an added asset to effective biological control strategies.

It is essential that biologic and genetic variation within and among species be characterized in order to develop effective biological control strategies. This chapter focuses on the importance of population genetic divergence and implications for biological control. Measures of genetic variation and genetic divergence which can be used to identify taxonomic entities at various stages of the evolutionary process are also important. Emphasis is placed on (1) the role of systematics and population genetics in implementation of effective biological control programs; (2) the strengths and limitations of experimental approaches commonly used to measure genetic differentiation; (3) the complexities of population structures, especially in the patterns of genetic variation; (4) the levels of genetic divergences among common taxonomic entities, such as geographical populations, races, biotypes, subspecies, semispecies, and sibling and nonsibling species; and (5) methods for improving analysis of population variability and structures.

Biological control is based on a premise of manipulating pest species, or natural enemies of a pest species, to reduce the pest's negative effects. In order to achieve successful pest suppression using biological control and to ensure that the suppression program has no adverse environmental effects, we believe information about naturally occurring genetic variation and the structure or organization of that variation, for all species involved, are critical elements to the program. In this chapter, we discuss these crucial elements by drawing on a variety of examples where documentation has been well established for a number of arthropod pests, primarily of medical and veterinary importance, and with which the authors have had some experience. We believe that the examples discussed here are applicable to the biological control of most arthropod pests.

II. ROLE OF SYSTEMATICS IN BIOLOGICAL CONTROL PROGRAMS

The literature on biological control contains examples of both successes and failures. The failure of many efforts can be traced to inadequate knowledge of biosystematics and population ecology.[3,4] In contrast to α-taxonomy based principally on morphology, biosystematics must be a multidisciplinary approach, including studies on population structure, behavior (including chemical cues such a pheromones), sound production, habitat preference, physiology, and population biology.[1,5] The systematics of both host(s) or prey(s) is required, as well as knowledge of microhabitats and host preferences.

Many parasitic species of arthropods have not been identified and named. Many of these are likely sibling species. The identification of these cryptic species is important because different parasitic species attack different hosts. If they attack the same host, this may impact on their effectiveness as biological control agents. For example, of the four known cryptic species of the hymenopteran species *Aphytis maculicornis,* only one, *A. paramaculicornis* (a sibling species of *A. maculicornis),* demonstrated effectiveness as a biological control agent of the olive scale.[1] In the genus *Aphytis,* Rosen and DeBach[6] identified 35 new species for biological control. *A. chrysomphali,* until recently considered a single species, was found to be a complex of sibling species. In fact, about 20% of the 100 known species of the genus *Aphytis* are now known to be groups of sibling species.[2] Identification of these species provides greater precision in the use of these parasites as biocontrol agents.

The majority of chalcidoid hymenopterans remain unnamed or misclassified. Less than 10% of an estimated 500,000 species of parasitic Hymenoptera have so far been described. This taxonomic order contains the most numerous and effective species of natural enemies of arthropod pests. The Ichneumonidae constitute about 20% of all known parasitic insects of biological control significance. Little or no biological information is available for about 97% if the Ichneumonidae species.[7] Using genetic markers, sibling species have been documented in almost every parasitic insect group that has been analyzed.

Misidentification of an exotic parasite species could lead one to assume that the species is already present in a region. This could result in ignoring its usefulness as a biological control resource. The inability to correctly understand genetic diversity in the form of detecting infraspecific taxa (races, strains, biotypes, ecotypes, subspecies, semispecies, etc.) can severely affect the outcome of the biological control program. Steiner discusses this in detail in Chapter 1. For example, the Japanese and Chinese races of the parasite *Compriella bifasciata,* when introduced into California, displayed distinct host (red scale) preferences.[1] Hybrids generated from the two races were not successful in attacking either host and disappeared from the parasite population. The two races existed sympatrically and are, perhaps, semispecies or species that are in the process of becoming reproductively isolated through selection in favor of homokaryotype matings. The absence of systematics studies has led to a delay in the biological control of several pests, including the sugarcane leafhopper in Hawaii, the coffee mealybug in Kenya, the California red scale, the walnut aphid, the olive scale in California, the Florida red scale in Hong Kong, the coconut scale in Fiji, as well as a number of other pests.

If an imported parasite was collected from a misidentified cryptic host, it may not have an appropriate host to attack when introduced into another region or country. Release of the wrong hymenopteran parasitoid braconid, *Chelonus texanus,* led to the failure of the biological control of the Karoo caterpillar, *Loxostege frustalis,* in South Africa. The correct parasitoid was late identified as *C. curvimaculatus.*[2]

The prudent approach to identifying unknown cryptic taxa is to designate a particular population as a neophyte to serve as a reference taxon. This population should be characterized in terms of its genetic makeup (chromosomal type, allozyme frequencies, and DNA variability patterns) and ecological, biological, and possibly physiological attributes. This information provides the basis to identify and characterize variant population structures and new taxonomic entities. Although genetic characterization of natural populations is important, i.e., for monitoring establishment and dispersal, it is not always directly crucial for the success of biological control. It is the host, habitat association, and the mating system that ultimately are critical under natural conditions.[8] It is easier to identify reproductive potential between sympatric taxa, but with allopatric taxa the only option is to test mating relationships in large artificial laboratory cages. These studies are laborious and time consuming, but the security of the biological control program is usually worth the investment.[1]

III. EXPERIMENTAL APPROACHES AND MEASURES OF GENETIC DIFFERENTIATION

The genetic differentiation between populations and species is the result of the interaction of random chance, mutation, migration, and natural selection. Consequently, natural populations are often differentiated from one another at different degrees. We believe knowledge of the patterns of genetic differentiation within all the relevant species is essential to ensure an effective control program. The patterns provide insight into the role of the major evolutionary factors in shaping the genetic structure of the species. These evolutionary factors are those which influence a species' response to a control program. All biological control strategies must be developed with some understanding of the effects of genetic drift, gene flow, and mutation and adaptation.

The methods commonly used to measure genetic variation, and hence differentiation between populations, include a comparative study of different banding patterns of polytene chromosomes, laboratory hybridization studies, isozyme electrophoresis, mitochondrial and ribosomal DNA restriction fragment length polymorphisms (mtDNA RFLP and rDNA RFLP), and methodologies using polymerase chain reaction (PCR) with a variety of different DNA markers.[9-12] The use of random amplified DNA polymorphisms generated by random amplified polymorphic DNA (RAPD)-PCR in taxonomic studies is discussed by Narang et al. in Chapter 3. PCR products of specific DNA regions can also be analyzed using restriction enzymes. Each method has strengths and limitations.

A. LABORATORY HYBRIDIZATION CROSSES

Studies of the laboratory hybridization between different populations provide a test of genetic affinities. Genetic divergence may produce results that range from complete reproductive incompatibility to complete fertility. Low affinity is indicated when hybrid eggs fail to hatch. A higher affinity is indicated when hybrid adults are produced. The degree of sterility of these adults determines the extent of gene flow possible between the parents.

The major weakness of hybridization measures of genetic affinity, especially when fertile progeny are produced, is that such results reveal little of what may happen in nature and where one or more isolating mechanisms may operate. Two taxa may mate in a laboratory cage, even though in nature there may be ecologic, geographic, or ethologic barriers to mating. Allopatric populations, in particular, present even greater difficulty in determining conspecific relationships. In natural populations, although hybridization between two species may occur, species status is justified as long as the gene exchange is sufficiently infrequent and each gene pool evolves independently.

B. CHROMOSOMAL DIFFERENTIATION

Genetic relationships based on chromosome structural differences require species which possess polytene chromosomes. Consequently, the majority of such studies of insect pests have been with species of *Anopheles* and *Sciara*.[13,14] Comparisons may be made using polytene band-for-band comparisons or by measuring regions of similarities (synapsis) and dissimilarities (asynapsis) of the chromosomes of interspecific hybrids. The degree of synapsis is assumed to be directly proportional to the genetic similarity between the parental species. Loops are formed in the areas of inversions, although small inversions may remain undetected. There is no evidence that the degree of synapsis or asynapsis in chromosomes of hybrids reveals the level of reproductive incompatibility between two taxa. Many well-defined species of *Anopheles* are homosequential and thus cannot be distinguished by this chromosomal method.[13,14] Another limitation is the varying degrees of synapsis of homologs in hybrids among taxa with similar banding patterns. Homologous chromosomes, which are separated in interspecies hybrids, indicate an accumulation of major differences in the genes responsible for chromosome pairing. The degree of chromosomal variation, observed for various taxonomic categories, is highly variable. Chromosomal inversions cover a wide array of genes of unknown or uncertain adaptive functions.[15] Inversions may include transposons and other genes responsible for chromosomal rearrangements. Thus, chromosomal polymorphism, though a good indicator of species variability, can be misinterpreted and lead to erroneous conclusions concerning the occurrence of taxonomic/evolutionary entities.

C. PROTEIN/ALLOZYME DIFFERENTIATION

The most information on population and taxa variation for systematics and population genetics has been obtained using electrophoretic analysis of enzyme variation. Allelic frequencies are well suited for statistical analysis to estimate levels of genetic variability and to identify differentiated taxa. Although identification of sympatric taxa may be straightforward, identifying allopatric samples, as distinct taxa based solely on significant differences in allele frequencies, can be misleading. For example, Murphy et al.[16] proposed that at least three unlinked, fixed, or nearly fixed electromorphs should be used an indication of differentiated taxa, but such identification should be confirmed by studies on ecology, behavior, and population biology of the population samples.

A useful measure of genic differentiation is the fixation index (F_{st}). F_{st} estimates the amount of allelic variation that is attributable to differentiation among subpopulations. Qualitatively, a range of 0.05 to 0.15 for the F_{st} indicates moderate differentiation, while a range of 0.15 to 0.25 indicates great differentiation. An F_{st} of more than 2.5 is indicative of very great differentiation.[17] Low values as small as 0.05 or less do not mean that differentiation is negligible.[18] The most commonly used measures of genetic divergence or genetic distance, such as D_N,[19,20] D_R,[21] D_{arc},[22] and the Manhattan distance, D_M,[18] are related to F_{st}.[17,23] In the case of allopatric taxa, significant genetic distance values can be indicative of cryptic taxa, especially if diagnostic loci are involved.

Electrophoretic methods to assess genetic differentiation have limitations.[17] First, allozyme loci are not representative of the entire genome. Second, for any enzyme locus, only a proportion of the actual number of protein variants may be detectable. Third, estimates of evolutionary levels of divergence between specific taxa, based on genetic distance values (D) or F_{st}, may differ between groups of species. Fourth, genetic distance values based on allozyme frequencies can vary in magnitude depending on the type of enzymes and the number of loci analyzed. The differences in the level of protein variation, such as represented in the average expected heterozygosity (H_e), in many cases, can depend on the choice of loci and is subject to large sampling variation.[12,24,25] The level of polymorphism, in general, is the highest in the hydrolytic and the least in glycolytic and Kreb's cycle enzymes.[26] Furthermore, the principle assumption required in calculations of D values from allozyme and DNA RFLP frequency data that the rates of gene substitutions per locus are uniform across all loci and lineages is unrealistic because it ignores well-documented effects such as drift, selection, and hitchhiking on polymorphisms.[10,25,27,28]

D. DNA MOLECULAR APPROACHES

DNA RFLP data can be used to estimate the extent of variation within and among taxa, level of gene flow, and effective population size. Restriction site data can be used to estimate genetic distance (d) based on the number of nucleotide substitutions between taxa X and Y after divergence from a common ancestor. The genetic distance d is equal to (ln S)/r, where S (proportion of unchanged restriction sites) is equal to $2n_{XY}/(n_X + n_Y)$ and n is the number of bases in the endonuclease recognition sequence.[29] The n_{XY} is the number of identical sites common between the two taxa, and n_X and n_Y are the total number of restriction sites in taxa X and Y, respectively.

Mitochondrial (mt) DNA RFLP data can be used for analysis of population structures and for tracing maternal lineages. The mitochondrial genome is maternally inherited and, in contrast to the nuclear genome, is free from the effect of sexual recombination.[30-32] As a result, mtDNA shows less intrapopulational and more interpopulational variation than autosomal nuclear loci. Therefore, mtDNA RFLP analysis is suited for detecting cryptic species, but less appropriate than allozymes for studying genetic structures of populations within species. Ribosomal (r) RNA gene restriction site variation has been used to define infraspecific taxa in *A. gambiae*.[33,34] Statistical procedures for the analysis of RFLP data can be found elsewhere.[17,25,35]

Restriction fragment length data, when used to estimate phylogenetic relationships among taxa, may result in misleading conclusions. Ideally, important phylogenetic characters have chromosomal map locations. Presence or absence of restriction sites at specific locations are considered character states and can be treated as alleles. However, because construction of restriction maps is time consuming, many inves-

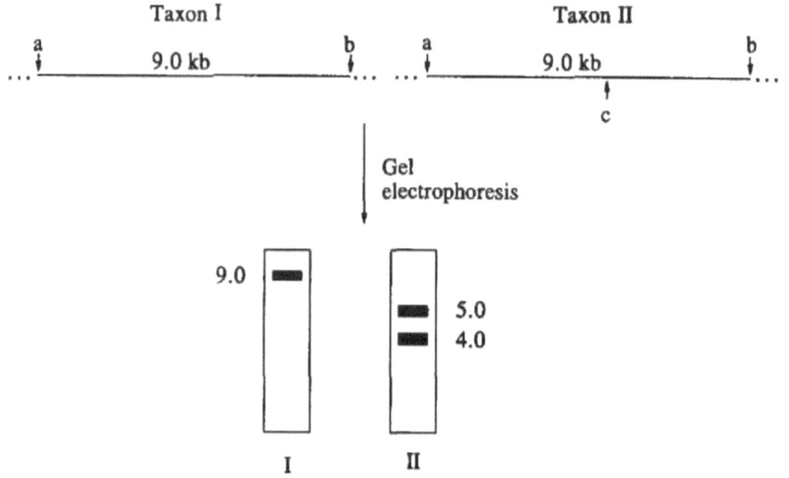

FIGURE 1. Example of expected fragments (size in kilobases) from digestion of a specific DNA region. Though two of the three sites (a, b, c) are common, between the two taxa they do not show any identical restriction fragments.

tigators use the restriction fragments themselves as the character states to estimate phylogenetic relationships. RFLP data as such lacks independence among these character states. For example, two taxa sharing two of the three restriction sites may have no restriction fragments in common (Figure 1), which would preclude the use of these RFLPs for phylogenetic analysis. Insertions and deletions between homologous sites on a restriction map in one, but not in the other taxon, could mask shared sites between the two taxa. In addition, even though a restriction site may be constructed, asymmetry in the probabilities of gaining or losing sites causes problems for phylogenetic analysis.[36]

The PCR is primarily a method of generating a virtually unlimited number of copies of a specific or nonspecific target DNA. The methods and applications of PCR can be found elsewhere in References 37 to 39. Several investigators have used polymorphic PCR-amplified DNA fragment patterns for taxonomic purposes.[40–42] Size differences in single amplified fragments have been used for identification of genetically differentiated taxa within the parasitic wasp *Trichograma*.[43] Narang et al. in Chapter 3 discuss this technique further.

Chemical characteristics such as cuticle hydrocarbons, alkaloid venom, and trail pheromones, analyzed by gas chromatography, have been used to define populations, races, and sibling species in social Hymenoptera.[44,45]

E. ESTIMATES OF GENETIC DIFFERENTIATION DETERMINED BY DIFFERENT APPROACHES

Although estimates of genetic differentiation among taxa from cross-mating relationships, chromosomal banding patterns, and genetic distance based on protein or DNA RFLPs provide useful information on species' genetic structures, each

TABLE 1
Incongruence in Estimates of Genetic Differentiation Based on Nei's[20] Genetic Distance (D) and Crossing Relationships[13] Among Species of *Anopheles maculipennis*[1]

Cross (*Anopheles*)		Pairwise species		Survival eggs	Fertility of F adults	
Female	Male	differentiation D	% Hatch	to adults	♀	♂
A. labranchiae	*A. atroparvus*	0.25	N	N	F	S
A. labranchiae	*A. maculipennis*	0.33	80	22	S	S
A. labranchiae	*A. messeae*	0.33	Low	0	—	—
A. labranchiae	*A. subalpinus*[a]	0.33	65	27	S	S
A. maculipennis	*A. atroparvus*	0.33	—	—	S	S
A. maculipennis	*A. messeae*	0.24	04	—	—	—
A. maculipennis	*A. subalpinus*	0.16	84	27	S	S
A. messeae	*A. atroparvus*	0.33	0–low	0	—	—
A. sacharovi	*A. atroparvus*	0.63	Low	Low	—	S
A. subalpinus	*A. atroparvus*	0.33	N	25	F:S (1:1)	S
A. subalpinus	*A. messeae*	0.20	60	60	—	S

Note: N, normal; F, fertile; and S, sterile.

[a] The species *subalpinus* is an egg variant phenotype of *A. melanoon*.

measure may be more or less influenced by different evolutionary processes. There may be little relationship between two or more measures of divergence for a particular species pair. For example, species pairs of *Anopheles*, with a larger genetic distance, may show greater hybridization than a pair which is genetically similar based on allozyme data. Apparently, in some species, allozyme differentiation is not indicative of reproductive incompatibility (Tables 1 to 3).

Allozyme variability in Hymenoptera is very low in most species.[46] However, considerable variability is evident from the RAPD-PCR analysis, as described by Narang et al. in Chapter 3. Little isozyme variation was observed in populations of the tephritid fruit fly, *Anastrepha fraterculus*, from Sào Paulo, Brazil. However, mtDNA RFLP analysis revealed high levels of variability, where haplotypes belong to one of two remarkably different major groups. This indicated a heterogeneous origin of these populations by introgression or hybridization.[47] The two cryptic taxa (C-1 and C-2), in the sibling species C of the *Anopheles quadrimaculatus* complex, revealed significant differentiation in allele frequencies at five loci,[48] yet little or no chromosomal or mtDNA RFLP differences are known.[48a,48b] In contrast, some insects such as the screwworm reveal high levels of intraspecific mtDNA variation.[49,50] Little or no mtDNA variability has been observed in aedines and anophelines mosquitoes,[51,52] monarch butterflies,[53] and the corn earworm *Helicoverpa zea*.[54]

A variety of techniques are available to address systematics issues ranging from intrapopulational relationships to species phylogeny. However, not all evolutionary problems can be solved by simply applying the latest technique. A new systematics

TABLE 2
Incongruence in Estimates of Genetic Differentiation (Nei's[20] Genetic Distance, D) and Crossing Relationships Among Species of *Anopheles gambiae* Complex[a]

Cross (*Anopheles*) Female sp.	Male sp.	Nei's D	% Hatch	Survival eggs to adults	Female[b]	Fertility of F_1 adults
A. gambiae	*A. arabiensis*		N	N	F	S
A. arabiensis	*A. gambiae*	[0.13]	N	N	F	S
A. gambiae	*A. quadriannulatus*		—	—	—	S
A. quadriannulatus	*A. gambiae*	[0.19]	—	—	—	S:F (9:1)
A. gambiae	*A. melas*		N	N	F	S
A. melas	*A. gambiae*	[0.24]	~50	sl	F	S
A. gambiae	*A. merus*		N	N	F	S
A. merus	*A. gambiae*	[0.44]	~50	sl	—	S
A. arabiensis	*A. quadriannulatus*		—	—	—	S
A. quadriannulatus	*A. arabiensis*	[0.14]	—	—	—	S
A. arabiensis	*A. melas*		~50	Low	F	S
A. melas	*A. arabiensis*	[0.10]	—	sl	F	S
A. arabiensis	*A. merus*		N	N	F	S
A. merus	*A. arabiensis*	[0.41]	~50	sl	—	S
A. quadriannulatus	*A. melas*		—	—	—	S
A. melas	*A. quadriannulatus*	[0.19]	—	—	—	S
A. quadriannulatus	*A. merus*		—	—	F	S
A. merus	*A. quadriannulatus*	[0.48]	—	—	F	S
A. melas	*A. merus*		N	<N	F	S
A. merus	*A. melas*	[0.41]	N	N	F	S
A. bwambae	*A. melas*		—	N	—	S
A. melas	*A. bwambae*	[0.31]	—	N	F	S
A. bwambae	*A. merus*		—	N	F	S
A. quadriannulatus	*A. bwambae*	[0.31]	—	—	F	—

Note: N, normal; F, fertile; and S, sterile.

[a] For detailed results of laboratory crosses refer to review by Narang and Seawright.[14]

[b] Hybrid females in most cases are less fertile than perental females; some lay significantly fewer eggs, especially F_1 females of *A. arabiensis* × *A. gambiae* and *A. gambiae* × *A. melas*.

technique may be inappropriate or less cost effective than a traditional technique. No one technique is superior to others under all circumstances. Hillis and Moritz[25] discussed the suitability of different techniques for systematics and stated, "Advances in technology often promote various kinds of chauvinism. ... Most recently, some proponents of sequencing have suggested that all other techniques are superfluous and outdated. ... Only by combining data from various morphological and molecular techniques is it possible to obtain a comprehensive view of evolution." A complete discussion on systematics research problems and the suitability of each of the different techniques can be found elsewhere in Hillis and Moritz.[55]

TABLE 3
Incongruence in Estimates of Genetic Differentiation Based on Nei's[20] Average Genetic Distance (D) and Crossing Relationships Among Species of *Anopheles quadrimaculatus* Complex[a]

Cross			Nei's	Mean %	%Survival egg	Fertility of F_1	
Female	×	Male	D	hatch[b]	to adults	Female	Male
A	×	B_1[c]	0.16	~N	<50[d]	SS	—
A	×	B_2[c]	0.44	74(70–80)	50(35–82)	SS	S
B	×	A	0.44	72(66–78)	71(70–78)	SS	S
A	×	C-1	0.44–0.61	<50	63 ± 27[e]	—	S
C-1	×	A	0.44–0.61	65(47–80)	24 ± 20[e]	—	S
B	×	C-1	0.61	54(20–90)	17(5–50)[f]	—	S
						SS	SS
						S	—
C-1	×	B	0.61	76(40–95)	30(10–50)	SS	SS
				75(58–92)	1(0–2)	S	—
C-1	×	C-2	0.08–0.02				

Note: SS, semisterile; S, sterile; and N = normal.

[a] Crossing relationship data from Lanzaro et al.,[86] Narang and Seawright,[14] and Kaiser;[139] genetic distances from Narang and Seawright.[14]
[b] Considerable intraspecific variation in geographical strains was observed.
[c] Two types of B were observed: B_1 was all-female producing and B_2 produced males below the normal 1:1 ratio. When B_1 and B_2 stocks were mexid, the inbred stock B (B_1B_2) males when crossed to A females produced near normal sex ratio. Isozyme differentiation was not detected.
[d] All males die in pupal stage.
[e] Female mortality in embryonic or larval stage.
[f] Crosses between B and C-1 produced three types of progeny: no males, 35% males, or all males and no females.

IV. COMPLEXITIES OF SPECIES POPULATION STRUCTURES

A. SPECIES

A "genetical" species is an assemblage of variable populations evolving as a collective unit and reproductively isolated from other such units. Natural populations are far more complex taxonomic entities than observed by most experimental methodologies.[17] Most species are widely distributed, inhabiting environmentally heterogeneous geographic areas. The total population of a species is usually subdivided into discrete subpopulations, with each population often having a characteristic genetic composition.[12,56]

The patterns of intra- and interspecific diversity varies considerably among different insect groups. Genetic diversity reveals geographic (latitude, longitude), clinal, temporal/seasonal, host-specific, and other ecologically related patterns in a number of pest species.[48,57-66]

In general, in most species, there seems to be some relationship between the geographic proximity among populations and the genetic distance among them. In others, such as in the beetle *Collops georgianus*, genetic distances among populations are more related to contiguity than to the geographic distance between them.[67,68]

B. ALLOZYME POLYMORPHISM AND INTRASPECIES GENETIC DIFFERENTIATION

Numerous studies have shown that levels of intraspecies genetic variation vary considerably between species and between populations within species. For example, the Hymenoptera are the least polymorphic for allozyme loci of all insect groups. The average heterozygosity (H) per species, a measure of genetic variation averaged over all loci analyzed, ranged from 0 to 0.8 in solitary, 0.02 to 0.08 in presocial, and 0 to 0.10 in eusocial species.[26,46,69] In contrast, H ranges from 0.11 to 0.31 in Coleoptera, 0 to 0.39 in Lepidoptera, and 0.02 to 0.24 in Diptera.[26]

Some isozyme loci are highly polymorphic in one group of insects, but monomorphic or nearly so in others. As shown in Figure 2, aldolase is highly polymorphic in *Eurylophella* mayflies, but monomorphic in eusocial hymenopterans and *Culex*. The α-glycerophosphate dehydrogenase (α-Gpdh) loci are polymorphic in most species (H = 0.198 in *Culex*), but low to monomorphic in eusocial insects (H = 0.043), aedine, and anopheline mosquitoes.[26,70,71] Phosphoglucomutase (*Pgm*) is highly polymorphic in most species (H = 0.717 in Coleoptera), but reveals low variability in the eusocial Hymenoptera (H = 0.010). Several other isozyme loci (*Pgi, Mdh, Ldh, Ald*) are highly polymorphic in presocial, but monomorphic or nearly so in eusocial Hymenoptera. Consequently, intraspecies average expected heterozygosities vary considerably in different insect groups (Figure 3). Steiner[26] discussed possible explanations for the differences in isozyme variability in different insect groups based on different evolutionary constraints in different insects or different biochemical strategies in response to environmental and biotic conditions. Although adaptive differences may explain different isozyme patterns, it is also likely that differences could be due to chance and/or differences in species history.[72]

Generally, the greater the genetic distance, the greater the time since the divergence of two species.[73] In insects, levels of genetic divergence are used to provide an indication of the age of the taxonomic group, and hence Nei's genetic difference (D) between various taxonomic entities varies considerably from group to group. For example, Nei's D values among nonsibling species of *Eurylophella* mayflies are lower than those among the majority of subspecies, sibling, and nonsibling species of different insect groups.[74] The genetic differentiation among conspecific populations of *Culicoides v. occidentalis*, D = 0.164,[63] is 164 times more than that found in the *Drosophila repleta* species group, D = 0.001.[75] Among different species, there is considerable variation in the magnitude of genetic differentiation between similar taxonomic entities. The average Nei's D among chromosomal races of *Anopheles albitarsis*[71] is ten times greater than among races of *D. sylvestris*.[76] It is about 80-fold greater among subspecies of *C. variipennis*[63] than among those of *Toxorhynchites rutilus*.[77] The divergence among *C. variipennis*, D = 0.316,[63] exceeds those among the majority of the sibling species of many other groups. Similarly, the average Nei's D values among many subspecies in *Drosophila*, *Culicoides*, and

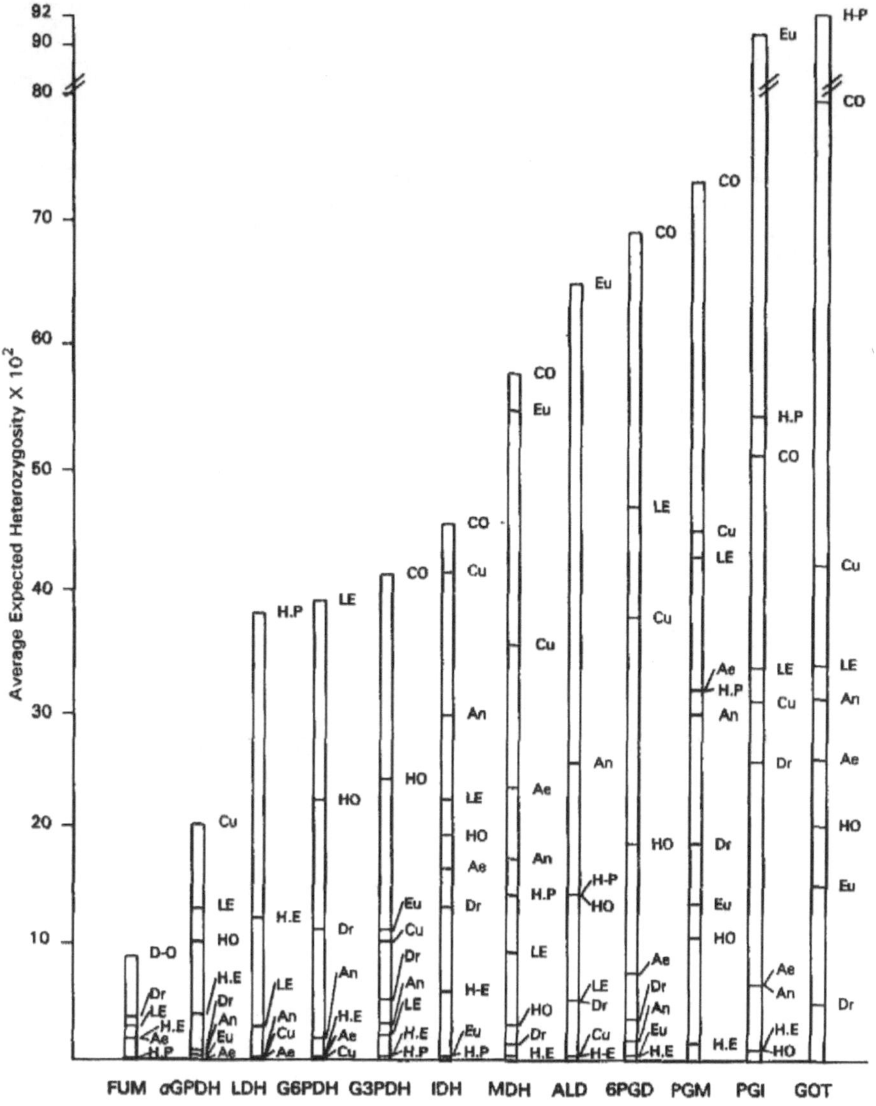

FIGURE 2. Relative levels of expected heterozygosity (H) values of twelve energy-related enzymes for selected insect orders and families (data from Steiner[26]). Ae = *Aedes*, An = *Anopheles*, CO = Coleoptera, Cu = *Culex*, Dr = *Drosophila*, D.O = other Diptera, Eu = *Eurylophella* mayfly, H.E = eusocial Hymenoptera, H.P = presocial Hymenoptera, HO = Homoptera, and LE = Lepidoptera.

Anopheles spp.[63,71,75] exceed those among nonsibling species of *Eurylophella* mayflies[74] and *Collops* beetles.[67,68]

The genetic distance between sibling species *A. sacharovi* and *A. atroparvus*[14] is about 130-fold greater than between *A. freeborni* and *A. hermsi*.[78] The level of divergence among sibling species in the genus *Anopheles* varies considerably, five-fold each in *A. maculipennis* and *A. gambiae* complexes and about eight fold in the

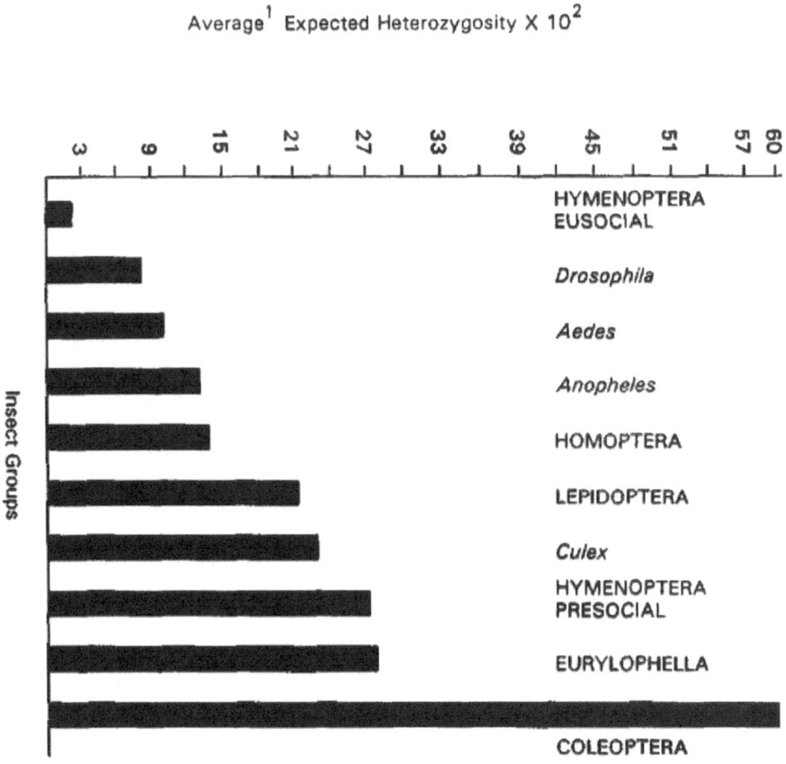

FIGURE 3. Average expected heterozygosity (H) levels in various insect groups (data from Steiner[26]). H values were based on variability in 12 energy-related proteins (ALD, FUM, G-6-PDH, G-3-PDH, GOT, α-GPDH, IDH, LDH, MDH, PGI, PGM, and 6-PGDH).

A. quadrimaculatus complex (Figure 4).[14,48] Therefore, the magnitude of genetic distance between any pair of populations cannot be used to designate the level of taxonomic relationship. There is no relationship between the levels of divergences and the evolutionary stages of speciation.

C. COMPLEXITIES IN PATTERNS OF GEOGRAPHIC VARIATION WITHIN SPECIES

Widely distributed species have complex patterns of genetic variation, which may differ between geographic regions. Geographic patterns of chromosomal and allozyme differentiation vary greatly in different species. In the beetle *C. georgianus*, there is less genetic variation among populations located in the center of the species range than populations observed at the edge of the species distribution.[67] Such a finding of differences between central and marginal populations is reminiscent of several species of *Drosophila*.[75, 79-81] In *D. willistoni* and *D. robusta*, genetic heterozygosity is lower in geographically peripheral and ecologically marginal populations than in the center of the range.[80] It is difficult to define marginal and central populations in

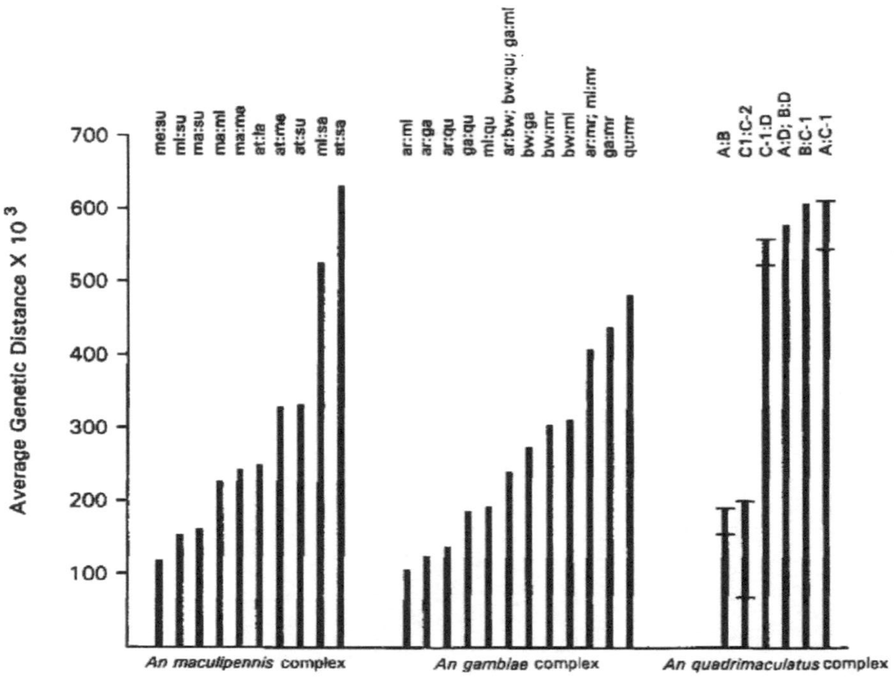

FIGURE 4. Comparisons of levels of genetic differentiation D[19,20] among sibling species pairs of three species complexes (from Europe, Africa, and North America) in the genus *Anopheles*.[14] *A. maculipennis:* la = *labranchiae*, ma = *maculipennis*, me = *messeae*, ml = *melanoon*, sa = *sacharovi*, su = *subalpinus; A. gambiae:* ar = *arabiensis*, bw = *bwambae*, ga = *gambiae*, and at = *atroparvus*.

many pest species; however, pest populations which have been continuously subject to suppression activities might be considered in an ecologically marginal environment when compared to populations which are not being suppressed. Many Caribbean island populations of the yellow fever mosquito, *Aedes aegypti*, have been subjected to chemical control, yet there is no clear relationship between the heterozygosity values of those "ecologically" marginal populations and uncontrolled populations elsewhere in the world.[62,66,82] Similarly, there is little evidence to support this pattern of genetic variation in *Culicoides variipennis*.[63,64]

The different aspects of the genome measured by various methods may be influenced by the same evolutionary factors, but to a different degree. Accordingly, the patterns of geographic variations observed using each method may differ. For example, *Anopheles albimanus* revealed higher genic variability in the northern than southern Colombia populations, which are separated by over 1000 km. However, these populations were homosequential and lacked inversion polymorphism.[83] Sibling species C-1 and C-2 of *A. quadrimaculatus* showed significant differentiation at 17% of 29 loci examined,[48] but had identical mtDNA RFLP patterns and showed little or no chromosomal differentiation.[48a,48b] The populations of *A. quadrimaculatus* species A and B reveal geographic and clinal patterns in allozymes,[84-86] whereas little or no differences were found in mtDNA restriction patterns.[52]

D. CHANGES IN GENETIC COMPOSITION

The genetic structure of populations within any species may change as a result of a variety of evolutionary factors. Any attempts at biological control must recognize the temporal instability of the genetic composition of the populations, the factors influencing this instability, and the effects and influences of this instability.

Random chance changes in allele frequencies, also termed genetic drift, are the result of the random sampling of gametes from one generation to another.[18] The importance of genetic drift and its effect of population differentiation, as well as the temporal genetic instability of populations, is sometimes greatly unappreciated. The magnitude of the sampling change in allele frequencies is directly related to the effective population size, N_e. The smaller the N_e, the greater the probability that the allele frequencies of the subsequent generation will differ from the previous generation due to chance alone. The smaller the N_e, the higher the probability of a large change.

The population structure of *Aedes aegypti* in the Caribbean is greatly influenced by genetic drift resulting directly from changes in population size due to intensive control vector. Wallis, Tabachnick, and Powell[66] reported extensive genetic differentiation among Caribbean island populations of *A. aegypti*, but little geographic pattern in allozyme gene frequencies. In contrast, Caribbean island populations of *D. willistoni* were genetically homogeneous.[87] Tabachnick and Wallis[72] attributed *A. aegypti* heterogeneity between islands to the fluctuations in population size resulting from sporadic *A. aegypti* control programs. Genetic drift or chance changes in allele frequency in small populations were believed to be the cause of the large amount of genetic differentiation.

A related and similar effect of genetic drift occurs when a small number of individuals generate or "found" a new population. This results in a "Founder Effect" because, just as in genetic drift, the small N_e of the founders results in random chance changes in allele frequencies, compared to the original population. Colorado populations of *C. variipennis* illustrate this effect. During the cold Colorado months, *C. variipennis* exist only as larvae in a few aquatic habitats. Each spring, new larval habitats are colonized by founders emerging as adults from the permanent habitats. The temporary sites show greater gene frequency variations and significant temporal variation compared to permanent sites, which is exactly as expected from a Founder Effect.[64]

Introduced pest species may exhibit reduced genetic variation in comparison to ancestral stocks as a result of founder events. In the U.S., the alfalfa weevil, *Hypera pestica* (Gyllenhal) is less variable when analyzed for morphological mutations compared to the weevil in Europe and the Middle East.[88] The introduction of the gypsy moth in the U.S. is the result of a founder event, which resulted in less genetic variation of U.S. populations compared to ancestral European gypsy moths.[89] The consequences of genetic drift and founder effects may have significant implications for biological control efforts. Species with the population structure of *C. variipennis* in Colorado require a control strategy different from *A. aegypti* in New Orleans. *A. aegypti* in New Orleans exhibit little genetic variation between geographic areas and were found to be panmictic or random breeding populations.[61] Hence, a biological control program to suppress *C. variipennis* would be more successful if targeted at the genetically stable overwintering larval populations. Successful control would be difficult during the summer when numerous temporary populations have been established that are genetically different. This problem does not exist with *A. aegypti* in New Orleans, and effective control would be more successful if directed at the adult mosquito. Consideration of the consequences of founder effects is important for the effectiveness of releasing and the subsequent survival of predators for insect biocontrol. As a first principle, the chances of successful introduction of a natural predator or parasite is greater if the released insect contains appropriate levels of genetic diversity comparable to ancestral natural populations. The release of mass-reared predator insects has often demonstrated the difficulties associated with the loss of genetic variation. Mass-reared screwworms, *Cochliomya hominivorax* (Coquerall), lost competitiveness in a sterile-insect release program with accompanying loss in genetic variation at an isozyme locus compared to wild flies.[90]

In contrast to genetic drift, which enhances differentiation, migration or gene flow prevents differentiation between populations. An understanding of the nature of gene flow between populations of pest species, as well as the nature of the gene flow between populations of insects used as biological control agents, is critical in developing effective control strategies. Species with low levels of gene flow do not contribute to an influx of new genes into the control region which might adversely affect control effectiveness. On the other hand, these same species will not easily result in effective control through natural migration of the control agents to new populations. Hence, estimates of gene flow are important.

The level of differentiation between populations, F_{st}, decreases as the number of migrants (Nm) increases, where approximately $F_{st} = 1/4Nm + 1$. A second method in estimating Nm from allele frequency data is by use of a method based on the frequencies of private or conditional alleles.[91,92] N is the population size, m is the migration rate (number of migrants/N), and Nm is the number of migrants irrespective of the population size. Nm estimates have been observed using both methods for several pest species. Populations of *Culicoides variipennis* in Colorado, within a 1- to 2-km radius, exhibit Nm values of 2 to 7 migrants between populations.[64] This level of gene flow is enough to prevent significant genetic differentiation between the populations, but the number of migrants is considered only a small proportion of the thousands of *C. variipennis* which can be found during the summer in any *C. variipennis* larval and adult site. On the other hand, estimates of Nm among treehoppers of the *Enchenopa binotata* complex had a value of only 0.181 among populations from the different host plants, *Ptelca*.[93] This low level of gene flow supports the large effect genetic drift has on populations of treehoppers from this host. In general, treehoppers from a variety of host plants revealed a low Nm value.

The migratory insect *Heliothis virescens* has a large Nm value of 125.[94,95] Very low levels of genic differentiations, suggesting a high Nm value, were observed in the Australian moths *H. armigera* and *H. punctigera*.[96] Additional studies of gene flow between populations of different pest species, as well as gene flow within potential predator species, are required to determine effects on biological control strategies.

Whereas the effects of genetic drift will be observed for many loci of a species, the effects of specific adaptations and natural selection on the genetic structure of a species occurs on selected loci. The details of types of selection and their effects on population structure are beyond the scope of this chapter. Numerous population genetics texts are available and should be consulted for further details. Suffice it to point out that a knowledge of environmental adaptive factors acting on populations is an important consideration to biological control efforts with natural enemies such as parasites and predators. Control strategies, which go against natural adaptations and the selective pressures on the relevant species, are likely to be ineffective. The alfalfa weevil *Hypera pestica* of the eastern U.S. can encapsulate eggs of the parasitoid *Brathyplectes onronlionis*, thus reducing its effectiveness as a biocontrol agent. This adaptation is not present in western populations, where control has been effective.[97] Demonstrating the action and nature of natural selection on specific genes is not a trivial task, as evidenced by the few available examples.

Some examples of adaptive genes include the acetylcholine esterase (ACHE) gene associated with malathion resistance in *D. melanogaster*;[98] ACHE and esterase (*Est*-6) in *D. melanogaster*;[99] *Est* allozymes correlated with insecticide usage in *Myzu persicae*;[100] and a phosphoglucoisomerase (*Pgi*) allele associated with flight capacity, survival, and mating competitiveness in *Colias* butterflies.[101,102] Steiner[26] has written an excellent review which presents evidence for the adaptive nature of allozyme polymorphisms. The mathematical formulas for calculations on the magnitude of different types of selections are discussed by Speiss.[103]

There are many examples of geographic variations within pest species that are of importance to the interaction between a pest insect and its predators. Important adaptive traits have been described for predaceous green lacewings in the *Chrysoperla carnea* species complex. Tauber and Tauber[104] discussed various adaptational differences among strains of this complex, particularly in terms of diapause, habitat association, and color. *Diabrotica* sp., serious pests of corn, exhibit egg diapause and feeding and dispersal adaptations to different crop rotations.[105] Scale insect populations possess adaptations which differ between local demes, allowing these insects to utilize different host species in different environments.[106] The fall armyworm, *Spodoptera frugiperda*, has been shown to differ in development and response to insecticides, as well as to allozyme frequencies between strains depending on the host crop.[107] Although the adaptive nature of variation is critically important to understanding interactions between insect pest and predator populations, there are few studies of adaptive significance of specific genes and genetic mechanisms. Greater understanding of genetic mechanisms will provide opportunities for the use of biological control strategies which interrupt adaptations through genetic and environmental modifications.

It is often difficult to determine where one population ends and another begins. Therefore, each population must be defined in terms of overall genetic composition or genetic structure. The population is defined in terms of the type, pattern, and extent of genetic variation or gene frequency distributions and the spatial and temporal changes that occur. The genetic makeup of a population changes as a result of many factors, i.e., mutation, natural selection, gene flow, genetic drift, or population interactions.

V. LEVELS OF GENETIC DIFFERENTIATIONS IN MOST COMMON INFRASPECIFIC TAXA

Poorly defined taxonomic terms, which are directly relevant to levels and patterns of divergence, have led to confusion in the literature. The terms local population, biotype, ecotype, races, varieties, strains, gamodemes, subspecies, semispecies, and sibling species have all been used to describe variations between organisms and groups of organisms. These common taxonomic terms are based on morphologic, ecologic, physiologic, behavioral, and mating patterns and, in a few cases, on genetic (chromosomal, allozymes) differences.

Several investigators have attempted to clarify this situation by characterizing the genetic structures of biotypes and host races, as well as to establish guidelines for classifying taxa into other subspecific taxonomic categories.[24,108-112] There are seemingly few distinctions between terms such as biotype and race. We believe that characterization of populations as a formal biotype or race is of questionable value. There are specific instances when a biotype trait may be important in pest control, but it is the trait which is of interest, not the formal classification of the biotype or race. The following subsections are brief discussions of the common taxonomic entities.

A. LOCAL POPULATION

The term local population, often used to describe a group of individuals of the same species living together in the same territory, leads to interpretative problems. The individuals may not be homogeneously distributed, but may be clustered as small family groups. The boundaries between local populations are not always clearly defined. Local populations are not always independent of one another, and their genetic makeup is subject to changes due to genetic drift and the migratory ability of some species.

B. MENDELIAN POPULATION

The need to define discrete territorial boundaries of local populations can be reduced by defining the Mendelian population. A Mendelian population is a community of interbreeding, sexually reproducing individuals with a potential Hardy-Weinberg equilibrium. For example, in many genetic population models, a population is usually considered large with random mating between individuals and a sex ratio of 1:0. All adults have an equal probability of producing offspring, and the generations are nonoverlapping. In nature, these populations rarely exist because populations are usually subjected to various evolutionary forces and undergo genetic changes.

C. NATURAL POPULATION

A natural population is defined by a number of attributes which include a finite number of breeding individuals, fluctuations in numbers over generations, fluctuations in sex ratio, variations in numbers of progeny, deviation from random mating within the population, overlapping generations, and the proportion of individuals that immigrate in each generation. The combined effect of these parameters results in the effective population size, N_e. Because of the complex interactions of these factors, it is difficult to determine the effective size in arthropod populations. The genetic divergence (Nei's D) among populations varies considerably in different species. It is 0.003 in *Anopheles quadrimaculatus* species C,[48] 0.163 in *Culicoides v. occidentalis*,[63] and 0.173 in *Toxorhynchitis ambionensis*[77] Similarly, Nei's D among geographic populations varies from species to species. It has a value of 0.008 in *A. albimanus*,[83] 0.143 between two populations of *A. aquasalis*,[113] 0 to 0.219 in *Collops georgianus*, and 0.013 to 0.691 in *C. tricolor*.[67,68] This variation in Nei's D from species to species is due to the diverse evolutionary factors, which also vary from species to species and contribute to why populations are genetically different.

D. RACE

A race is composed of groups of individuals or populations which differ from other groups within formally recognizable subspecies by virtue of distinct allozyme frequencies, features of chromosome structure, or some biological characteristic. Races may or may not be distinguishable morphologically. Races of the same species are both interfertile and produce fertile offspring. Populations with some degree of genetic differentiation (as measured by chromosomal rearrangements; the fixation index, F_{st}; or genetic distance, D), may be considered as races.

Races can be allopatric or sympatric. *Geographic races* occupy different geographic subdivisions of the range of a species. *Ecologic races*, or ecotypes, are confined to a particular restricted habitat. *Physiologic races* reveal physiological differences rather than morphological character differences. *Chromosomal races* differ in chromosome structure (cytotype) or in chromosome number. *Host race* shows a preference for a specific host plant or insect that differs from other races.[114] The level of genetic divergence among races varies considerably. It is 0.046 to 0.051 between two races of *A. albimanus*[83] and 0.089 to 0.397 among three chromosomal races of *A. albitarsis*.[71]

E. BIOTYPE

The biotype concept is discussed in detail by Steiner in Chapter 1. Biotypes differ from one another in attributes other than host preference. Biotypes may differ in insecticide resistance or life history and habitat attributes. Like races, consistent gene frequency differences for a number of loci may be observed in samples of different biotypes. Allozyme differences have been reported between biotypes of corn leaf aphids,[115] apple maggot fly,[24,116] and the grape phylloxera *Daktulosphatra viifoliae*.[117] However, Wellso et al.[118] were unable to detect electrophoretic differences among biotypes of the hessian fly. Biotype-specific enzymes[119] and mtDNA polymorphisms[120] have been reported in the U.S. greenbug aphid *Schizaphis graminum* populations. Narang et al.[120a] used RAPDs to distinguish biotypes of a parasitic hymenopteran. Some host-adapted biotypes of *S. graminum* have been differentiated by RAPD-PCR patterns.[40]

F. SUBSPECIES

Subspecies are considered to represent the first stage of speciation. It is a geographic subdivision of a species which shows consistent morphologic or biologic differences from other parts of the species range.[121] The distinction between race and subspecies may be argued. Races which differ in easily observable characteristics are often given Latin names and formally recognized as subspecies.[17] The term subspecies is sometimes limited to cases where there is a degree of reproductive isolation between population divisions, whereas race refers to those subdivisions in which there is no barrier to reproduction, but only geographic divisions. Subspecies which produce sterile male hybrids in one or both directions represent incipient reproductive isolation with no behavioral or sexual isolation.

Sympatric subspecies of *Aedes aegypti*, occupying different habitats in East Africa, differ morphologically, behaviorally, and in allozyme frequencies. The existence of a region with reduced gene flow between subspecies might lead falsely to conclusions of distinct species. However, since the subspecies of *A. aegypti* also occur with gene flow in West Africa, species status does not seem valid.[62] Regions where subspecies are sympatric are particularly instructive for determining estimates of gene flow and the validity of subspecies status. At times, *Culex pipiens* subspecies have been considered different species. Tabachnick and Powell[122] analyzed populations where *C. pipiens* subspecies are sympatric and identified stable, freely interbreeding populations of *C. pipiens* in the central valley of California. Extensive gene flow was found between *C. pipiens pipiens* and *C. pipiens quinquefasciatis* in these populations because individuals were identified with morphology of one subspecies, but allozymes of the other. These are clearly subspecies with differences in mating behavior, human biting preferences, and the ability to lay eggs without a blood meal.

As noted above, the level of genetic differentiation among subspecies of different species vary so widely that, for practical purposes, these values, by themselves, are of no taxonomic value. For example, in spite of apparent differences in morphological, physiological, and behavioral traits among populations of *A. aegypti* throughout the world, intraspecific differentiation is low. Of the two subspecies, the darkly colored "sylvan" *A. aegypti formosus* (Walker) is the predominant form in Africa south of Sahara, and the lightly colored domestic "type form" *A. aegypti aegypti* (L) prevalent outside Africa, is divided into several "genetic-geographic" groups. The genetic distance, D, between *A. aegypti formosus* and various genetic geographic groups of the subspecies *A. aegypti aegypti* ranges from 0.0193 to 0.0762.[62,65] This is only one third to one fifth the amount of distance between the two subspecies in each of *Drosophila willistoni* and *D. equinoxialis.*[123] Again, this suggests the relatively recent divergence of the *A. aegypti* subspecies from one another compared to *D. willistoni* subspecies.

Nei's D among subspecies in other insect groups are 0.004 to 0.042 in *T. rutilus*,[77] 0.103 to 0.249 in *Anopheles aquasalis*,[71] 0.128 in *D. repleta* to 0.230 in *D. willistoni* species group,[75] 0.217 in *C. pipiens*,[124] and 0.185 to 0.316 in *Culicoides variipennis*.[63]

G. SEMISPECIES

Semispecies represent a second stage in the process of speciation. Sympatric populations may show strong ethological isolation and sterility barriers. Laboratory crosses between populations may produce sterile male F_1 hybrids in both directions. Semispecies are considered incipient species. Populations of different semispecies may be identified only by hybridization experiments and sometimes by chromosomal arrangements. Semispecies show greater morphologic differences and greater genetic differences than are usually found in subspecies and races. Gene exchange is not as high as in subspecies, but it is higher than between species. The semispecies concept could be of significance to biological control efforts if mating barriers which could interfere with the control effort are present. An imported biological control agent may attack one semispecies and not the other. Different subspecific taxa of parasitoids and predators might show differences in their effectiveness to destroy different semispecies of host pests.

H. SIBLING SPECIES

Sibling species represent a third stage of speciation. Sibling species are morphologically indistinguishable, but reproductively isolated from one another, and are thus a true biological species. Speciation is complete. The degree of sterility among sibling species of *A. maculipennis, A. gambiae, and A. quadrimaculatus* groups varies considerably with the strain and direction of the crossing relationship (Tables 1 to 3).

Sibling species in most species complexes generally show fixed allelic differences or diagnostic loci. However, the proportion of fixed differences can vary greatly in different groups and even among sibling species of the same species complex. Some examples of proportions of diagnostic loci among sibling species in different species complexes can be found in Figure 5.

The level of differentiation between sibling species varies considerably from group to group. There is a wide variation (about 155-fold) in genetic distance among

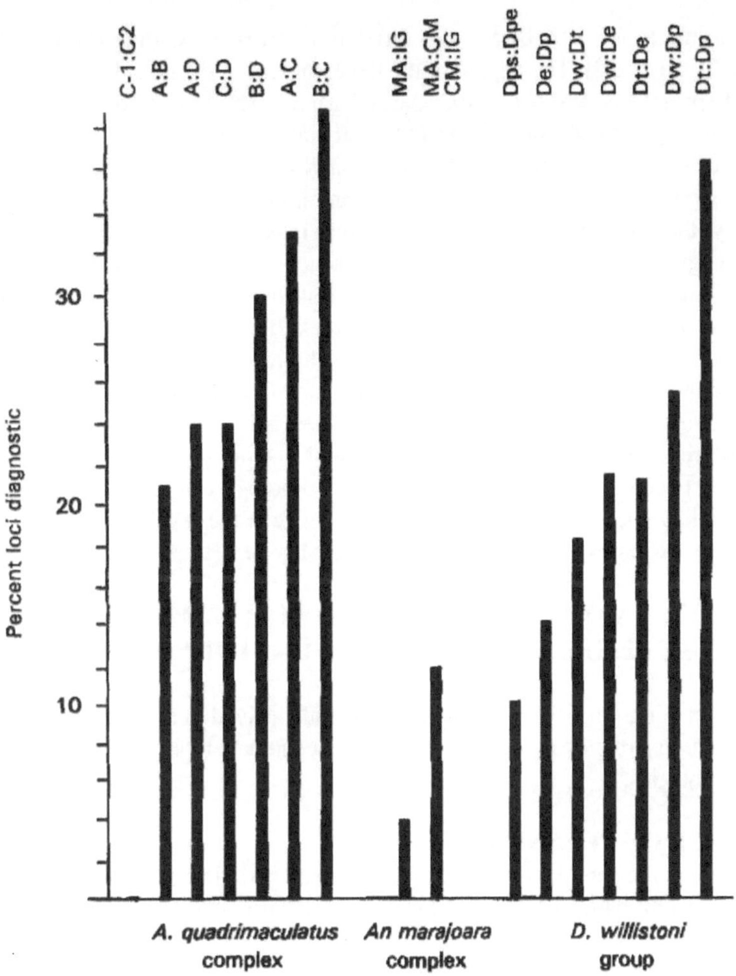

FIGURE 5. Percentage of diagnostic loci among pairwise sibling species. Number of loci examined: 33 in sibling species, A, B, C-1, C-2, and D of the *A. quadrimaculatus*;[14,48,48a] and 25 in *A. marajoara* complex.[125] Data on *Drosophila* was derived from Ayala.[87] De = *D. equinoxialis*, DP = *D. paulistorun*, Dpe = *D. persimillis*, Dps = *D. pseudoobscura*, Dt = *D. tropicalis*, and Dw = *D. willistoni*. MA = from island of Marajo; IG = Iguape in the state of Sào Paulo in Brazil.

the sibling species of *Anopheles*. Nei's D has a value of 0.005 in the *A. freeborni* complex,[78] 0.080 to 0.243 in the *A. albitarsis* complex,[125] and 0.772 (range 0.592 to 0.772) in the *A. darlingi* complex.[71,126] In three well-characterized anopheline species complexes (Figure 4), D varies from 0.071 to 0.368 in the *A. gambiae* complex, 0.080 to 0.614 in the *A. quadrimaculatus* complex, 0.119 to 0.633 in the *A. maculipennis* complex,[14,48,127] and 0.480 between sibling species *Aedes hendersoni* and *A. triseriatus*.[128] These D ranges overlap those estimated for many subspecies.

A comparative analysis of cross-mating relationships among sibling species shows wide variation from complete sterility to partial fertility of hybrid males. The aberrations in the testes of hybrid males differ in degrees, depending on the types of crosses (Table 1 to 3). The degree of the development of the gonads varied with the parental species and the sex of the species' parent. The fertility of hybrid females ranged from partial fertility of the hybrid females to all females being sterile, a mixture of sterile and fertile females, and finally to all female hybrids being fertile. Sex-ratio variation in hybrid progeny is common in crosses among sibling species. Similarly, there is a great variation in hatching rates and hybrid larval mortality.[14]

The degree of chromosomal asynapsis is not always an indication of sibling species status. Even homosequential species fail to show synaptic hybrid chromosomes. In many sibling species, most genetic differentiation is limited to the X chromosomes. The autosomes are largely synaptic, and the X chromosomes are asynaptic in the hybrids of *Anopheles gambiae* and *A. arabiensis* and those of the *A. maculipennis* complex. In others, both autosomes and X chromosomes have undergone differentiation as indicated by asynapsis.[14,75] Although there is no direct relationship between the degree of chromosomal homology and reproductive isolation among sibling species, there is a correlation between the level of asynapsis in the hybrid chromosomes and the depression in fertility of the hybrid female.

VI. SAMPLING METHODS FOR IMPROVING CHARACTERIZATION OF POPULATIONS STRUCTURES

The following guidelines are applicable for identification of taxa and for analyzing population structures of species of importance in developing effective biological control strategies.

A. CHOICE OF TECHNIQUES

Because of the complexity of patterns of genetic variation in field populations and the analytical limitation of any single technique, the use of a combination of different techniques becomes desirable. All of the techniques reviewed in this chapter may provide required information, but each has its limitations. Hillis and Moritz[25] discussed parameters for choosing the appropriate techniques for specific problems.

B. SAMPLE SIZE

There is no one optimal sample size adequate for sensitive scrutiny of all possible types of population structures. The smaller the difference in allele frequencies between two taxa, the larger the sample sized necessary to detect real differences. One can estimate the number of diploid individuals required to detect given differences in allele frequency. The number of specimens required for estimation of genetic distances among taxa depends on the amount of polymorphism.[129] Larger samples decrease variance of the sample means and provide better population estimates.

Most studies on intraspecific variation and population genetics require examination of a large number of population samples for a large number of loci. It is advisable to conduct a pilot study of a few selected populations rather than an ambitious study on population structuring, species boundaries, and phylogenetic reconstruction all at once.[25] Empirical guidance can be obtained from numerous published reports. For example, analysis of about 100 individuals were necessary to identify host races and microgeographic variation in the apple maggot fly.[130,131] Analysis of the malic enzyme *(Me)* locus in 100 to 200 individuals led to the identification of microgeographic genetic subdivisions of the anopheline sibling species *A. quadrimaculatus* species C.[48] It is interesting that many studies may analyze too many individuals per population. The same information may be obtained with fewer individuals by analyzing more populations and more loci to obtain a better indication of species genetic structure.

C. NUMBER OF LOCI

There are no strict guidelines for determining adequate numbers of loci or restriction enzymes for identifying all possible population subdivisions. Generally, the more loci used, the greater the likelihood of estimating differentiation. For example, increasing the number of gene-enzyme systems for the analysis of the population structure of *Rhagoletis pomonella* led to an increase in the number of polymorphic loci from 9 in 1976[116] to 17 in 1988.[132] If the goal is to identify loci associated with host adaptation or linked to other important characteristics, then analysis of several hundred loci would be desirable, depending on the total size of the species genome. Though techniques for histochemical staining of over 100 enzymes are available, cost can be a limiting factor. Studies may routinely analyze variability at about 35 to 40 enzymes yielding data on about 50 loci. The ideal solution would be to complement allozyme studies with DNA RFLP and PCR data.[11] Closely related species, diverged within the past five million years, are best studied by examining fast-evolving isozyme loci.[25]

D. LIFE HISTORY STAGES

Conclusions about genetic structuring can be greatly influenced by the developmental stage analyzed. Generally, researchers use a particular stage based on the ease of sampling or electrophoretic resolution of some isozymes. Most information on host races comes from analyzing matings obtained by sampling individuals actually mating and ovipositing on the host plants. To understand how and when selection causes differentiation, it is important to sample different stages. This is best illustrated by studies on microgeographic variation in the apple maggot fly.[130] Allozyme analysis of adults raised from larvae collected from infested fruits from hawthorn trees revealed significant intertree allele frequency differences. This intertree genetic heterogeneity could be due to either disruptive selection limited to the larval stage, with random mating among adults, or nonrandom mating among adult subpopulations associated with host trees. Genetic variation in mountain pine beetles associated with the host tree is subject to similar questions.[133]

Another example of genetic differentiation, where sampling of all stages is warranted, is with *A. quadrimaculatus* species C.[48] Species C has been found principally near the Gulf Coast, in river and spring systems of northwest Florida, and the

Ogeechee River in Georgia. Samples from the Gulf Coast revealed significant allelic differences from those collected in the river system in the Florida panhandle, indicating pronounced population substructuring. Mitochondrial DNA RFLP studies, using up to 35 restriction endonucleases, did not reveal differences between samples from these two sites. Examination of the intermediate region in Florida led to identification of a boundary zone between the Gulf and river populations near Porter's Lake in Florida.[133a] These results confirm that discrete populations separated by a discrete boundary occur in Florida. Allozyme loci may be under different evolutionary pressures than the mtDNA markers. Sampling of larval stages will be necessary to explain genic differentiation between these taxa.

E. SAMPLING TO DISTINGUISH GEOGRAPHIC AND HOST-RELATED VARIATION

Inadequate geographic sampling can confound geographic variation with host-related variation. This problem is greatest in allopatric sampling of putative biotypes, i.e., if samples of one biotype are collected from one geographic area and those of another biotype are collected from another area. Geographic variation can easily be confused with host-related variation. We have discussed before that geographic variation generally shows an inverse relationship with the degree of dispersal or migratory behavior. It ranges from very low in migratory species to very high in insects with limited mobility. In addition, microgeographic variation can occur.[40,67,68,130,131,134] Therefore, conclusions based on allopatric sampling should be avoided unless the two hosts do not share the same geographic area. In that case, variation between the two samples should be complemented with data from behavioral and ecological studies.

Sympatric sampling is informative, but microgeographic/habitat heterogeneity should be taken into consideration. For example, the pattern of mating inferred from allozyme frequency differences might be due to differences in habitats (dry upland vs. wet bottomland) rather than differences in the host plant. Unless putative races occur in the same habitat, the conclusion that the populations represent host races is tentative. These studies must again be supplemented with behavioral and ecological work.

The best evidence for host races is provided if populations showing genetic differentiation are collected from the area where the two host plants are intermixed or are within the known dispersal distance of the insects of the two host plants. In this case, allozyme or DNA RFLP or RAPD-PCR differences between samples of adult insects from the two plants would strongly suggest the existence of host races. However, these differences should be reproducible in microgeographic samples from other areas of the range of host plants.

F. SEASONAL AND TEMPORAL SAMPLING

Seasonal changes in frequencies of chromosome inversions in *Drosophila* are well documented.[135] Similar changes in molecular polymorphisms are also known.[136] Willis et al.[137] reported seasonal changes in the relative frequencies of cryptic taxa in the *A. quadrimaculatus* complex. Therefore, variation between samples of two putative races collected during different seasons can be confused with climate-related changes. The ideal solution to this problem would be to collect samples during the time when members of both host-associated races are actively mating and ovipositing. If this is not possible, conclusions of host races must be supported by the behavioral and ecological differences. Efforts should be made to repeat these genetic studies over several years for the conclusive establishment of host races.

VII. CONCLUSION

Genetic variation in natural populations is produced by mutations and genetic recombination and is organized, maintained, eliminated, and/or dispersed among subpopulations as a result of migration, natural selection, and genetic drift. These are the same factors which directly influence biological control agents and the manner in which pest species may respond to these agents. Accordingly, such information is critical to an effective biological control program. Geographic patterns of gene variation between species are complex. Generalizations to explain all patterns are difficult.

The degree of genetic divergence among taxa can be estimated by using laboratory hybridization and chromosomal, biochemical, and molecular methodologies. Not all levels of taxa comparisons involve the same amount or type of genetic differentiation, even when reproductive isolation arises through the process of geographic speciation. The time frame elapsed since speciation occurred is a primary determinant of the degree of genetic differentiation. Regulatory genes may play a primary role in the development of reproductive isolation and morphological evolution.[75,81,138] Morphologic divergence and loss of hybridization potential do not necessarily show parallel divergence in other genetic markers.

In this chapter, we have focused on techniques to measure arthropod population genetic structure and implications for developing effective biological control strategies. The examples of genetic analyses of arthropod populations are heavily biased toward studies of genetic markers, many of which are probably unrelated to important arthropod traits. We hope the reader has been convinced that information about genetic structure, obtained through the analysis of genetic markers, is applicable and relevant to important arthropod pests in terms of effective biological control programs. The future challenge will be to assess the genetic variability of arthropod pests and their biocontrol agents (parasites and predators) for traits of direct relevance to the pest and the biological control agent. This will require identification and the ability to monitor genes controlling behavior, host utilization, physiology, and host preference, as well as life history characteristics and the relationship of these factors to parasite/predator host preference.

An understanding of pest and natural enemy interactions in terms of population genetics has been an increasing goal of biological control specialists. Application of genetic methods for improving biocontrol agents in the many characteristics contributing to their effectiveness as biotic regulators of pest populations should offer great opportunities in the control of harmful pest species. Numerous geographical and biological "races" of parasitic species exist in nature. They differ in their potential value as biological control agents, and there are also marked differences in host adaptation between races. Several approaches in the application of genetics to improving biocontrol agents involve the selection of stocks of the maximum possible genetic variability for release in the specific environment. Reliance is then on the forces of natural selection to develop an effective strain. The development of races or strains in the laboratory that can operate effectively in environments unsuited to existing strains may be possible through such techniques as hybridization, irradiation, gene transfer, or other means of enriching the gene pool. To date, however, the transfer of specific genes into insects, whether pests or predators, and the incorporation of these genes into field populations is in its infancy. All of these ambitious undertakings should help achieve the goal of more effectively controlling arthropod pests without adverse impacts on the environment.

REFERENCES

1. **DeBach, P. and Rosen, D., Eds.,** *Biological Control By Natural Enemies,* Cambridge University Press, New York, 1991.
2. **Paterson, H. E. H.,** The recognition of cryptic species among economically important insects, in *Heliothis: Research Methods and Prospects,* Zalucki, M. P., Ed., Springer-Verlag, Berlin, 1991, 1.
3. **Delucchi, V., Rosen, D., and Salinger, E. I.,** Relationship of systematics to biological control, in *Theory and Practice of Biological Control,* Huffaker, C. B. and Messenger, P. S., Eds., Plenum Press, New York, 1976, 81.
4. **Rosen, D.,** The role of taxonomy in biological control program, *Agric. Ecosyst. Environ.,* 15, 121, 1986.
5. **Henry, C.,** *Chrysoperla johnsoni* (Neuroptera: Chrysopidae): acoustic evidence for full species status, *Ann. Entomol. Soc. Am.,* 86, 14, 1993.
6. **Rosen, D. and DeBach, P.,** *Species of Aphytis of the World (Hymenoptera: Aphelinidae),* Israel Universities Press, Jerusalem, and Dr. W. Junk, the Hague, 1979.
7. **Townes, H.,** Ichneumonidae as biological control agents, *Tall Timbers Con. Ecol. Anim. Contr. Habit. Mgmt.,* 3, 235, 1971.
8. **Paterson, H. E. H.,** The recognition concept of species, in *Species and Speciation,* Vrba, E., Ed., Transvaal Museum, Transvaal Museum Monograph, Pretoria, 1985, 4.
9. **Harris, H. and Hopkinson, D. A.,** *Handbook of Enzyme Electrophoresis on Human Genetics,* North-Holland Publishing, Amsterdam, 1976.
10. **Hillis, D. M.,** Misuse and modification of Nei's genetic distance, *Syst. Zool.,* 33, 238, 1984.
11. **Narang, S. K.,** Molecular approaches for the identification of population structures of economically important insects, in *The Eukaryotic Chromosome: Structure and Functional Aspects,* Sobti, R. C. and Obe, G., Eds., Springers-Norossa Publication Co., New Delhi, 1993, 59.
12. **Richardson, B. J., Baverstock, P. R., and Adams, M.,** *Allozyme Electrophoresis. A Handbook for Animal Systematics and Populations Studies,* Academic Press, Sydney, 1986.
13. **Kitzmiller, J. B., Frizzi, G., and Baker, R. H.,** Evolution and speciation within the Maculipennis complex of the genus *Anopheles,* in *Genetics of Insect Vectors of Disease,* Wright, J. and Pal, R., Eds., Elsevier, New York, 1967, 151.

14. **Narang, S. and Seawright, J. A.,** Genetic differentiation among members of species complexes in anopheline mosquitoes (Diptera: Culicidae), in *The Eukaryotic Chromosome: Structural and Functional Aspects*, Sobti, R. C. and Obe, G., Eds., Springers-Norossa Publication Co., New Delhi, 1990, 59.
15. **Prakash, S. and Levitan, M.,** Association of alleles of the *Esterase*-1 locus with gene arrangements of the left arm of the second chromosome in *Drosophila robusta*, *Genetics*, 75, 371, 1973.
16. **Murphy, R. W., Sites, J. W., Jr., Buth, D. G., and Haufler, C. H.,** Protein I: isozyme electrophoresis, in *Molecular Systematics*, Hillis, D. M. and Moritz, C., Eds., Sinauer Associates, Inc., Sunderland, MA, 1990, 45.
17. **Hartl, D. L. and Clark, A. G.,** *Principles of Population Genetics*, Sinauer Associates, Inc., Sunderland, MA, 1989.
18. **Wright, S.,** *Evolution and Genetics of Populations. Vol. 4. Variability Within and Among Populations*, University of Chicago Press, Chicago, 1978.
19. **Nei, M.,** Genetic distance between populations, *Am. Nat.*, 106, 283, 1972.
20. **Nei, M.,** Estimation of average heterozygosity and genetic distance from a small number of individuals, *Genetics*, 89, 583, 1978.
21. **Rogers, J. S.,** Measures of genetic similarity and genetic distance. Studies in genetics. VII, *Univ. Tex. Public.*, 213, 145, 1972.
22. **Cavalli-Sforza, L. L. and Edwards, A. W. F.,** Phylogenetic analysis. Models and estimation procedures, *Am. J. Hum. Genet.*, 19, 233, 1967.
23. **Swofford, D. L. and Selander, R. B.,** Biosys-1. A FORTRAN program for comprehensive analysis of electrophoretic data in population genetics and systematics, *J. Hered.*, 72, 281, 1981.
24. **Berlocher, S. H.,** THe complexities of host races and some suggestions for their identifications by enzyme electrophoresis, in *Electrophoretic Studies on Agricultural Pests, Vol. 39, Systematics Association*, Loxdale, H. D. and Hollander, J. D., Eds., Clarendon Press, Oxford, 1989, 51.
25. **Hillis, D. M. and Moritz, C.,** An overview of applications of molecular systematics, in *Molecular Systematics*, Hillis, D. M. and Moritz, C., Eds., Sinauer Associates, Inc., Sunderland, MA, 1990, 51.
26. **Steiner, W. W. M.,** Genetic differentiation and geographic distribution in insects: patterns and inferences, in *Evolution in Pest Insects: The Patterns of Variation*, Kim, K. C. and McPheron, B. A., Eds., Environmental Science & Technology Series, John Wiley & Sons, New York, 1993, 27.
27. **Birley, A. J.,** Drift, selection, hitchhiking, and DNA polymorphism in Drosophila, *Int. Congr. Genet.*, 17, 154, 1993.
28. **Braverman, J. M.,** A computer simulation model of hitchhiking with an allele, *Int. Congr. Genet.*, 17, 153, 1993.
29. **Nei, M. and Li, W. H.,** Mathematical model for studying genetic variation in terms of restriction endonuclease, *Proc. Natl. Acad. Sci. U.S.A.*, 76, 5269, 1979.
30. **Avise, J. C., Lansman, R. A., and Shade, R. O.,** The use of restriction endonucleases to measure mitochondrial DNA sequence relatedness in natural populations. I. Population structure and evolution in the genus *Peromyscus*, *Genetics*, 92, 279, 1979.
31. **Brown, W. M.,** The mitochondrial genome of animals, in *Molecular Evolutionary Genetics*, MacIntyre, R. J., Ed., Plenum Press, New York, 1985, 95.
32. **Wilson, A. C., Cann, R. L., Carr, S. M., George, M., Gyllenstein, U. B., Helmbychowski, K. M., Higuchi, R. G., Palumbi, S. R., Prager, E. M., Sage, R. D., and Stoneking, M.,** Mitochondrial DNA and two perspectives on evolutionary genetics, *Biol. J. Linn. Soc.*, 26, 375, 1985.
33. **Collins, F. C., Mendez, M. A., Rasmussen, M. O., Mehaffey, P. C., Bessansky, N. J., and Finnerty, V.,** A ribosomal RNA gene probe differentiates member species of the *Anopheles gambiae* complex, *Am. J. Trop. Med. Hyg.*, 37, 37, 1987.
34. **Finnerty, V.,** Ribosomal DNA probes for identification of member species of the *Anopheles gambiae* complex, *Fla. Entomol.*, 71, 288, 1988.

35. **Dowling, T. E., Moritz, C., and Palmer, J. D.,** Nucleic acids II: restriction site analysis, in *Molecular Systematics*, Hillis, D. M. and Moritz, C., Eds., Sinauer Associates, Inc., Sunderland, MA, 1990, 250.
36. **DeBry, R. W. and Shade, N. A.,** Cladistic analysis of restriction endonuclease cleavage maps within a maximum-likelihood framework. *Syst. Zool.*, 34, 21, 1985.
37. **Arnheim, N. and Erlich, H.,** Polymerase chain reaction strategy, *Annu. Rev. Biochem.*, 61, 131, 1992.
38. **Innis, M. A., Gelfand, D. H., Sninsky, J. J., and White, T. J., Eds.,** *PCR Protocols: A Guide to Methods and Applications*, Academic Press, San Diego, 1990.
39. **Simon, C., Franke, A., and Martin, A.,** The polymerase chain reaction: DNA extraction and amplification, in *Molecular Taxonomy*, Vol. H57, Hewitt, G. M., Ed., NATO Advanced Studies Institute, Springer-Verlag, Berlin, 1991, 329.
40. **Black, W. C., IV, DuTeau, N. M., Puterka, G. T., Nechols, J. R., and Pettorini, J. M.,** Use of the random amplified polymorphic DNA polymerase chain reaction (RAPD-PCR) to detect DNA polymorphism in aphids, *Bull. Entomol. Res.*, 82, 151, 1992.
41. **Kambhampati, S., Black, W. C., and Rai, K. S.,** Random polymorphic DNA of mosquito species and populations (Diptera: Culidae). Techniques, statistical analysis, and applications, *J. Med. Entomol.*, 29, 939, 1992.
42. **Raich, T., Archer, J. L., Robertson, M. A., Tabachnick, W. J., and Beaty, B. J.,** Polymerase chain reaction approaches to *Culicoides* (Diptera: Ceratopogonidae) identification, *J. Med. Entomol.*, 30, 228, 1993.
43. **Orrego, C. and Agudello-Silva, F.,** Genetic variation in the parasitoid wasp, *Trichograma* revealed by DNA amplification of the nuclear ribosomal repeat, *Fla. Entomol.*, 76, 519, 1993.
44. **Carlson, D. A.,** Hydrocarbons for the identification and phenetic comparisons: cockroaches, honey bees, and tsetse flies, *Fla. Entomol.*, 71, 333, 1988.
45. **Vander Meer, R. K. and Lofgrin, C. S.,** Use of chemical characteristics in defining populations of ants, *Solenopsis saevissima* complex, (Hymenoptera: Formicidae), *Fla. Entomol.*, 71, 323, 1988.
46. **Graur, D.,** Gene diversity in Hymenoptera, *Evolution*, 39, 190, 1985.
47. **Santos, P. D. and Matioli, S. R.,** Intraspecific mitochondrial DNA variation in the fruit fly *Anastrepha fraterculus* (Diptera: Tephritidae), *Int. Congr. Genet.*, 17, 153, 1993.
48. **Narang, S. K., Seawright, J. A., and Kaiser, P. E.,** Evidence for microgeographic genetic subdivision of *Anopheles quadrimaculatus* species C, *J. Am. Mosq. Control Assoc.*, 6, 179, 1990.
48a. **Narang, S. K.,** unpublished data, 1990.
48b. **Kaiser, P. and Mitchell, S.,** personal communication, 1993.
49. **Narang, S. K., Degrugillier, M. E., Krueger, C. M., and Taylor, D. E.,** Monitoring dispersal of screwworm populations by mitochondrial DNA and PCR, *Genet. Suppl.*, 11, 47, 1992.
50. **Roehrdanz, R. L.,** Intraspecific genetic variability in mitochondrial DNA of the screwworm fly (Cochliomyia: hominivorax), *Biochem. Genet.*, 27, 551, 1989.
51. **Kambhampati, S. and Rai, K. S.,** Variation in mitochondrial DNA of *Aedes* species (Diptera: Culicidae), *Evolution*, 45, 120, 1991.
52. **Mitchell, S. E., Narang, S. K., Cockburn, A. F., Seawright, J. A., and Goldenthal, M.,** Mitochondrial and ribosomal DNA variation among members of the *Anopheles quadrimaculatus* (Diptera: Culicidae) species complex, *Genome*, 35, 939, 1992.
53. **Brower, A. V. Z. and Boyce, T. M.,** Mitochondrial DNA variation in monarch butterflies, *Evolution*, 45, 1281, 1991.
54. **Narang, S. K., Degrugillier, M. E., and Lopez, J.,** Restriction fragment length polymorphism (RFLP) analysis of U.S.A. and Mexican populations of the corn earworm, *Helicoverpa zea*, Hagedorn, H., Hildebrand, J., Kidwel, M., and Law, J., Eds., *Proc. Second Int. Symp. Mol. Insect Sci.*, 1993, 116.
55. **Hillis, D. M. and Moritz, C., Eds.,** *Molecular Systematics*, Sinauer Associates, Inc., Sunderland, MA, 1990.
56. **Hedrick, P. W.,** Genetic polymorphism in heterogeneous environments. A decade later, *Annu. Rev. Ecol. Syst.*, 17, 535, 1986.

57. **Archie, J., Simon, C., and Wartenberg, D.,** Geographical patterns and population structure in periodical cicadas based on spatial patterns of allozyme frequencies, *Evolution*, 39, 1261, 1985.
58. **Carrio, R. Mendoza, A. H. D., and Latorre, A.,** Genetic variability detected by RAPDs-PCR in two aphid species, *Int. Congr. Genet.*, 17, 151, 1993.
59. **McDonald, I. C., Krysa, J. L., and Johnson, O. A.,** Genetic variation within and among geographic populations of *Diabrotica barberi* (Coleoptera: Chrysomelidae), *Ann. Entomol. Soc. Am.*, 78, 271, 1985.
60. **Nevo, E.,** Genetic variation in natural populations. Patterns and theory, *Theor. Popul. Biol.*, 13, 121, 1978.
61. **Tabachnick, W. J.,** Geographic and temporal patterns of genetic variation of *Aedes aegypti* in New Orleans, *Am. J. Trop. Med. Hyg.*, 31, 849, 1981.
62. **Tabachnick, W. J.,** Evolutionary genetican and arthropod-borne disease: the yellow fever mosquito, *Am. Entomol.*, 37, 14, 1991.
63. **Tabachnick, W. J.,** Genetic differentiation among populations of *Culicoides variipennis* (Diptera: Ceratopogonidae), the North American vector of bluetongue virus, *Ann. Entomol. Soc. Am.*, 85, 140, 1992.
64. **Tabachnick, W. J.,** Microgeographic and temporal genetic variation in populations of the bluetongue virus vector *Culicoides variipennis* (Diptera: Certaopgonidae), *J. Med. Entomol.*, 29, 384, 1992.
65. **Wallis, G. P., Tabachnick, W. J., and Powell, J. R.,** Macrogeographic genetic variation in a human commensal: *Aedes aegypti*, the yellow fever mosquito, *Genet. Res.*, 41, 241, 1983.
66. **Wallis, G. P., Tabachnick, W. J., and Powell, J. R.,** Genetic heterogeneity among Caribbean populations of *Aedes aegypti*, *Am. J. Trop. Med. Hyg.*, 33, 492, 1984.
67. **King, P. S.,** Macro- and microgeographic structure of a spatially subdivided beetle species in nature, *Evolution*, 41, 401, 1987.
68. **King, P. S.,** Distribution and genetic structure of two allopatric beetle (Coleoptera: Melyridae) species on rock outcrops in the southeast, *Ann. Entomol. Soc. Am.*, 81, 890, 1988.
69. **Wood, P. E. and Guttman, S. I.,** Genetic variation in *Neodiprion* (Hymenoptera: Symphyta: Dipronidae) sawflies and a comment on low levels of genetic diversity within the Hymenoptera, *Ann. Entomol. Soc. Am.*, 80, 590, 1987.
70. **Steiner, W. W. M., Tabachnick, W. J., Rai, K. S., and Narang, S. K., Eds.,** *Recent Developments in the Genetics of Insect Disease Vectors*, Stipes Publishing Co., Urbana, IL, 1982.
71. **Steiner, W. W. M., Narang, K. S., Kitzmiller, J. B., and Swofford, D. L.,** Genetic divergence in New World *Anopheles*, in *Recent Developments in the Genetics of Insect Disease Vectors*, Stipes Publishing Co., Urbana, IL, 1982, 63.
72. **Tabachnick, W. J. and Wallis, G. P.,** Genetic structure of *Aedes aegypti* populations in the Caribbean: ecological considerations, in *Ecology of Mosquitoes: Proceedings of a Workshop*, Lounibos, L. P., Rey, J. R., and Frank, J. F., Eds., Entomological Society of America Publications, College Park, MD, 1985, 371.
73. **Nei, M.,** *Molecular Population Genetics and Evolution*, American Elsevier, New York, 1975.
74. **Funk, D. H., Sweeney, B. W., and Vannote, R. L.,** Electrophoretic study of eastern North American *Eurylophella* (Ephemeroptera: Ephemerellidae) with the discovery of morphologically cryptic species, *Ann. Entomol. Soc. Am.*, 81 174, 1988.
75. **Ayala, F. J.,** *Molecular Evolution*, Sinauer Associates, Inc., Sunderland, MA, 1976.
76. **Carson, H. L.,** Sexual selection and speciation, in *Evolutionary Process and Theory*, Karlin, S. and Nevo, E., Eds., Academic Press, New York, 1986.
77. **Hilbrun, L. R., Willis, N. L., and Seawright, J. A.,** An electrophoretic comparison of laboratory colonies and natural populations of four species of *Toxorhynchitis* (Diptera: Culicidae), *J. Med. Entomol.*, 21, 336, 1984.
78. **Fritz, G. N., Narang, S. K., Kline, D. L., Seawright, J. A., Washino, R. K., Porter, R. C., and Collins, F. H.,** Diagnostic characterization of *Anopheles freeborni* and *An. hermsi* by hybrid crosses, frequencies of polytene X chromosomes and rDNA restriction enzyme fragments, *J. Am. Mosq. Control Assoc.*, 7, 198, 1991.

79. **Lewontin, R. C.,** *The Genetic Basis of Evolutionary Change,* Columbia University Press, New York, 1974.
80. **Prakas, S.,** Patterns of gene variation in central and marginal populations of *Drosophila robusta. Genetics,* 75, 347, 1973.
81. **Ayala, F. J.,** *Population and Evolutionary Genetics,* Benjamin/Cummings, Menlo Park, CA, 1982, 268.
82. **Wallis, G. P. and Tabachnick, W. J.,** Genetic analysis of rock hole and domestic *Aedes aegypti* on the Caribbean island of Anguilla, *J. Am. Mosq. Control Assoc.,* 6, 625, 1990.
83. **Narang, S. K., Seawright, J. A., and Suarez, M. F.,** Genetic structure of natural populations of *Anopheles albimanus* in Columbia, *J. Am. Mosq. Control Assoc.,* 7, 437, 1991.
84. **Narang, S. K., Kaiser, P. E., and Seawright, J. A.,** Dichotomous electrophoretic taxonomic key for the identification of sibling species A, B, and C of the *Anopheles quadrimaculatus* complex (Diptera: Culicidae), *J. Med., Entomol.,* 26, 94, 1989.
85. 84. **Narang, S. K., Toniolo, S. R., Seawright, J. A., and Kaiser, P. E.,** Genetic differentiation among sibling species A, B, and C of *Anopheles quadrimaculatus* complex (Diptera: Culicidae), *Ann. Entomol. Soc. Am.,* 82, 508, 1989.
86. **Lanzaro, G. C., Narang, S. K., and Seawright, J. A.,** Speciation in an anopheline (Diptera: Culicidae) mosquito: enzyme polymorphism and the genetic structure of populations. *Ann. Entomol. Soc. Am.,* 83, 578, 1990.
87. **Ayala, F. J., Powell, J. R., Tracey, M. L., Mourao, C. A., and Perez-Salas, S.,** Enzyme variability in the *Drosophila willistoni* group. IV. Genetic variation in natural populations of *Drosophila willistoni. Genetics,* 70, 113, 1972.
88. **Hsiao, T. H.,** Geographic and genetic variation among alfalfa weevil strains, in *Evolution of Insect Pests: Patterns of Variation,* Kim, K. C. and McPheron, B. A., Eds., John Wiley & Sons, New York, 1993, 311.
89. **Harrison, R. G., Wintermeyer, S. F., and Odell, T. M.,** Patterns of genetic variation within and among gypsy moths, *Lymantia dispar* (Lepidoptera: Lymantrildae), populations, *Ann. Entomol. Soc. Am.,* 76, 652, 1983.
90. **Bush, G. L.,** Planning a rational quality control program for the screwworm fly, in *The Screwworm Problem,* Richardson, R. H., Ed., University of Texas Press, Austin, 1978, 37.
91. **Slatkin, M.,** Gene flow in natural populations, *Annu. Rev. Ecol. Syst.,* 16, 393, 1985.
92. **Slatkin, M.,** Gene flow and the geographical structure of natural populations, *Science,* 236, 787, 1987.
93. **Guttman, S. I. and Weigt, L. E.,** Macrogeographic genetic variation in the *Enchenopa binotata* complex (Homoptera: Membracidae), *Ann. Entomol. Soc. Am.,* 82, 156, 1990.
94. **Korman, A. K., Mallet, J., Goodenough, J. L., Graves, G. B., Hays, J. L., Hendericks, D. E., Luttrell, R., Pair, S. D., and Wall, M.,** Population structure in *Heliothis virescens* (Lepidoptera: Noctuidae). An estimate of gene flow, *Ann. Entomol. Soc. Am.,* 86, 182, 1993.
95. **Mallet, J., Korman, A. K., Heckel, D. G., and King, P.,** Biochemical genetics of *Heliothis* and *Helicoverpa* (Lepidoptera: Noctuidae) and evidence for a founder event in *Helicoverpa zea, Ann. Entomol. Soc. Am.,* 86, 189, 1993.
96. **Daly, J. C.,** The use of electrophoretic data in a study of gene flow in the pest species *Heliothis armigera* (Hubner) and *H. punctigera* Wallengren (Lepidoptera: Noctuidae) in *Electrophoretic Studies on Agricultural Pests, Systematics Association, Vol. 39,* Loxdale, H. D. and Hollander, J. D., Eds., Clarendon Press, Oxford, 1989, 115.
97. **Puttler, B.,** Interrelationship of *Hypera postica* (Coleoptera: Curculionidae) and *Bathyplactes curonlionis* (Hymenoptera: Ichneumonidae) in the eastern United States with particular reference to encapsulation of parasite eggs by the weevil larvae, *Ann. Entomol. Soc. Am.,* 60, 1031, 1967.
98. **Morton, R. A. and Singh, R. S.,** The association between malathion and acetylcholinesterase in *Drosophila melanogaster, Biochem. Genet.,* 20, 179, 1982.
99. **Salam, A., El-Abidin, A., and Pinsker, W.,** Effects of selection for resistance to organophosphorus insecticides on two esterase loci in *Drosophila melanogaster, Genetica,* 55, 11, 1981.
100. **Brooks, C. P. and Loxdale, H. D.,** Survey of enzyme variation in British populations of *Muzus persicae* (Sulzer) (Hemiptera: Aphididae) on crops and weed hosts, *Bull. Entomol. Res.,* 77, 83, 1987.

101. **Watt, E. B.,** Bioenergetics and evolutionary genetics: opportunities for new synthesis, *Am. Nat.*, 125, 118, 1985.
102. **Watt, E. B., Carter, P. A., and Donohue, K.,** Females' choice of "good genotypes" as mates is promoted by an insect mating system, *Science*, 233, 1187, 1986.
103. **Speiss, E. B.,** *Genes in Populations*, 2nd ed., John Wiley & Sons, New York, 1989.
104. **Tauber, M. J. and Tauber, C. A.,** Adaptations to temporal variation in habitats: categorizing, predicting and influencing their evolution in agroecosystems, in *Evolution of Insect Pests: Patterns of Variation*, Kim, K. C. and McPheron, B. A., Eds., John Wiley & Sons, New York, 1993, 103.
105. **Kryson, J. L.,** Adaptations of Diabrotica to habitat manipulations, in *Evolution of Insect Pests: Patterns of Variation*, Kim, K. C. and McPheron, B. A., Eds., John Wiley & Sons, New York, 1993, 361.
106. **Hanks, L. M. and Denno, R. F.,** The role of demic adaptation in colonization and spread of scale insect populations, in *Evolution of Insect Pests: Patterns of Variation*, Kim, K. C. and McPheron, B. A., Eds., John Wiley & Sons, New York, 1993, 393.
107. **Pashley, D. P.,** Causes of host associated variation in insect herbivores: an example from fall armyworm, in *Evolution of Insect Pests: Patterns of Variation*, Kim, K. C. and McPheron, B. A., Eds., John Wiley & Sons, New York, 1993, 351.
108. **Daly, J. C.,** Methods for studying the genetics of populations of *Heliothis*, in *Heliothis: Research Methods and Prospects*, Zalucki, M. P., Ed., Springer-Verlag, Berlin, 1991, 157.
109. **Diehl, S. R. and Bush, G. L.,** Host shifts, genetic models of sympatric speciation and the origin of parasitic insect species, in *Proc. 5th Int. Symp. Insect-Plant Relationships*, Pudoc Wageningen, 1982, 297.
110. **Diehl, S. R. and Bush, G. L.,** An evolutionary and applied perspective of insect biotypes, *Annu. Rev. Entomol.*, 29, 471, 1984.
111. **Futuyma, D. J. and Peterson, S. C.,** Genetic variation in the use of resources by insects, *Annu. Rev. Entomol.*, 30, 217, 1985.
112. **Pashley, D. P.,** Current status of fall armyworm host strains, *Fla. Entomol.*, 71, 227, 1988.
113. **Narang, S. K., Kitzmiller, J. B., Galler, R., Rios, R. I., and Narang, N.,** Genetica de populacoes de anofelinos. III. Analise eletroforetica de *Anopheles aquasalis* (Diptera: Culicidae), *Rev. Brazil Pesq. Med. Biol.*, 12, 303, 1979.
114. **Bush, G. L.,** Sympatric host race formation and speciation in frugivorous flies of the genus *Rhagolitis* (Diptera: Tephritidae), *Evolution*, 23, 237, 1969.
115. **Steiner, W. W. M., Voegtlin, D. J., and Irwin, M. E.,** Genetic differentiation and its bearing on migration in North American populations of the corn leaf aphid, *Rhopalosiphum maidis*, (Fitch) (Homoptera: Aphididae), *Ann. Entomol. Soc. Am.*, 78, 518, 1985.
116. **Berlocher, S. H. and Bush, G. L.,** An electrophoretic analysis of *Rhagolitis* (Diptera: Tephritidae) phylogeny, *Syst. Zool.*, 31, 136, 1982.
117. **Williams, R. N. and Shambaugh, G. F.,** Grape phylloxera (Homoptera: Phylloxeridae) biotypes confirmed by electrophoresis and host susceptibility, *Ann. Entomol. Soc. Am.*, 81, 1, 1987.
118. **Wellso, S. G., Howard, D. J., Adams, J. L., and Arnold, J.,** Electrophoretic monomorphism in six biotypes and two populations of the hessian fly (Diptera: Cecidomyiidae), *Ann. Entomol. Soc. Am.*, 81, 50, 1988.
119. **Beregovoy, W. H. and Starks, K. J.,** Enzyme pattern in biotypes of the greenbug, *Schizaphis graminum* (Rondani) (Homoptera: Apididae), *J. Kans. Entomol. Soc.*, 59, 517, 1986.
120. **Martinez, D. and Castro, A.,** Greenbug aggressiveness variability within and between biotypes, *Int. Congr. Genet.*, 17, 155, 1993.
120a. **Narang, S. K., et al.,** unpublished results, 1993.
121. **Mayr, E.,** *Animal Species and Evolution*, Harvard Belknap Press, Cambridge, MA, 1963.
122. **Tabachnick, W. J. and Powell, J. R.,** Genetic analysis of *Culex pipiens* populations in the Central Valley of California, *Ann. Entomol. Soc. Am.*, 76, 715, 1983.
123. **Ayala, F. J., Tracey, M. L., Barr, L. G., McDonald, J. F., and Perez-Salas, S.,** Genetic variation in natural populations of five *Drosophila* species and hypothesis of the selective neutrality of protein polymorphism, *Genetics*, 77, 383, 1974.

124. **Cheng, M. L., Hacker, C. S., Pryor, S. C., Ferrell, R. E., and Kitto, G. B.,** The ecological genetics of the *Culex pipiens* complex in North America, in *Recent Developments in the Genetics of Insect Disease Vectors,* Steiner, W. W. M., Tabachnick, W. J., Rai, K. S., and Narang, S. K., Eds., Stipes Publishing Co., Urbana, IL, 1982, 581.

125. **Narang, S. K., Klein, T. A., Perera, O. P., Lima, J. B., and Tang, A. T.,** Genetic evidence for the existence of cryptic species in *Anopheles albitarsis* complex in Brazil. Allozyme and mitochondrial DNA RFLP patterns, *Biochem. Genet.,* 31, 97, 1993.

126. **Narang, S. K., Santos, J. M., Garcia, J., Kristakau, C. H., and Narang, N.,** Genetica de populacoes de anofelinos. IV. Estudos electroforetics das populacoes naturais de *Anopheles nuneztovari e Anopheles darlingi.* Correlacae genetica entre especies, *Acta Amazonica,* 9, 529, 1979.

127. **Narang, S. K., Kaiser, P. E., and Seawright, J. A.,** Identification of species D, a new member of the *Anopheles quadrimaculatus* species complex. A biochemical key, *J. Am. Mosq. Control Assoc.,* 5,317,1989.

128. **Mathews, T. E. and Munstermann, L. E.,** Genetic diversity and differentiation in northern populations of the tree-hole mosquito, *Aedes hendersoni,* (Diptera: Culicidae), *Ann. Entomol. Soc. Am.,* 76, 1005, 1983.

129. **Baverstock, P. R. and Moritz, C.,** Sampling design, in *Molecular Systematics,* Hillis, D. M. and Moritz, C., Eds., Sinauer Associates, Inc., Sunderland, MA, 1990, 13.

130. **McPheron, B. A., Smith, D. C., and Berlocher, S. H.,** Microgeographic genetic variation in the apple maggot, *Rhagoletis pomonella, Genetics,* 119, 445, 1988.

131. **McPheron, B. A., Smith, D. C., and Berlocher, S. H.,** Genetic differences between host races of *Rhagoletis pomonella, Nature (London),* 336, 64, 1988.

132. **Feder, J. L., Chilcote, C. A., and Bush, G. L.,** Genetic differentiation between sympatric host races of the apple maggot fly, *Rhagoletis pomonella, Nature (London),* 336, 61, 1988.

133. **Sturgeon, K. B. and Mitton, J. B.,** Allozyme and morphological differentiation of mountain pine beetles, *Dendroctonus pondersae* Hopkins (Coleoptera: Scolytidae) associated with host tree, *Evolution,* 40, 290, 1986.

133a. **Kaiser, P.,** personal communication, 1991.

134. **Pamilo, P.,** Genetic differentiation within subdivided populations of *Formica* ants, *Evolution,* 37, 1010, 1983.

135. **Dobzhansky, Th.,** Genetics of natural populations: IX. Temporal changes in the composition of *Drosophila pseudoobscura, Genetics,* 28, 162, 1943.

136. **Templeton, A. R. and Johnston, J. S.,** Life history evolution under plieotropy and K-section in a natural population of *Drosophila mercatorum,* in *Ecological Genetics and Evolution: The Cactus-Yeast-Drosophila Model System,* Barker, J. S. F. and Starmer, W. J., Eds., Academic Press, Australia, 1982, 25.

137. **Willis, N. L., Narang, S. K., Toniolo, S. R., and Shetty, N. J.,** Electrophoretic analysis of temporal changes in frequency distribution of sympatric sibling species of *Anopheles quadrimaculatus* (Diptera: Culicidae), *Genetics,* 116(Suppl.), 45, 1987.

138. **Lewontin, R. C.,** Population genetics, *Annu. Rev. Genet.,* 19, 81, 1985.

Chapter 3

DICHOTOMOUS RAPD-PCR KEY FOR IDENTIFICATION OF FOUR SPECIES OF PARASITIC HYMENOPTERA

S. K. Narang, R. A. Leopold, C. M. Krueger, and J. D. DeVault

TABLE OF CONTENTS

0-8493-2607-9/94/$0.00+$.50

I. INTRODUCTION

During the past decade, research on traditional biological control and its application to pest management has become increasingly attractive as an alternative to the use of synthetic organic pesticides. Many strains of exotic parasitic Hymenoptera and predatory Diptera and Coccinellidae are being investigated for their potential to control the Russian wheat aphid (RWA), *Diuraphis noxia*. There are no completely reliable methods for distinguishing various exotic strains of these biocontrol agents from those indigenous to the U.S. Release of unidentified and uncharacterized strains would make it impossible to document their establishment and dispersal. Therefore, genetic typing of strains prior to their release in the field is necessary.

The most common genetic methods for species and strain identifications include chromosomal polymorphism, protein electrophoresis, and, more recently, mitochondrial DNA restriction[1-4] patterns. Parasitic Hymenopterans, because of their very small sizes, are not easily analyzed by these techniques. Besides, low protein polymorphism has been reported throughout the Hymenoptera.[5] DNA polymorphisms detected by the polymerase chain reaction (PCR) technique are being widely used in fingerprinting genomes and in taxonomy.[6-10] Williams et al.[11] developed the random amplified polymorphic DNA (RAPD) technique for amplification of arbitrary regions of the genome using single 10-basepair (bp) oligonucleotide primers. Various regions of genomic DNA flanked by a pair of 10-bp priming sites within approximately 3.0 to 5.0 kb of each other are amplified. The RAPD-PCR method has been successfully applied in taxonomic and population variability studies in a number of insects including grasshoppers;[12] four species of aphids: *Schizaphis graminum*, *Acyrthosiphon pisum*, *Uroleucon ambrosiae*, and *D. noxia*;[13] two species of whitefly;[14] moths;[15,16] aedine mosquitoes of the *Aedes scutellaris* subgroup,[17] *A. aegypti*,[18] Hessian flies; deltocephaline leafhoppers; and tetranychid mites.[13]

Our overall objective is to identify DNA-based genetic markers for use in the identification and characterization of native and exotic parasites and predators. However, first, it is important to test the potential of the RAPD-PCR technique at the species level, especially to determine whether there exist species-specific patterns in geographically widespread populations of each species. In this chapter, we demonstrate DNA polymorphisms detected by RAPD-PCR in four species of parasitic Hymenoptera and present an easy-to-use dichotomous key for their identification.

II. MATERIALS AND METHODS

A. PARASITIC HYMENOPTERA

Adult insects from laboratory-reared strains of four species of Hymenoptera were provided by the National Biological Control Laboratory, U.S. Department of Agriculture, Animal and Plant Inspection Service (APHIS), Niles, MI. The sources and identification codes are shown in Table 1.

B. DNA ISOLATION

DNA was extracted from individual parasites by using a procedure similar to that of Simon, Franke, and Martin.[19] Individuals were homogenized in 52 μl of a lysis

TABLE 1
Strains of Four Species of Parasitic Hymenoptera Used for RAPD Analysis

Species	Geographical origin (collection, location)	Import ID	Species code
Aphidius matricariae	Moldavia, Kishinev	T90067	Am 1
	Pakistan, Northwest Frontier Province	T90003	Am 2
Diaeretiella rapae	Iran, Fars	T91050	Dr 1
	Iran, Ehglid	T90051	Dr 2
	Kazakhstan, Dmitrievka	T91089	Dr 3
	Pakistan, Parachinar	T91037	Dr 4
	China, Wuqia, Koshi	T91084	Dr 5
	Uzbekistan, Tashkent	T91150	Dr 6
	Syria, Qatara	T90026	Dr 7
Aphelinus asychis	Pakistan, Parachinar	T91062	Aa 1
	Kazakhstan, Dmitrievka	T91061	Aa 2
	China, Chillan	T90116	Aa 3
Aphelinus sp. nr *varipes*	Caucasus, Budannovak	T91086	Av 1
	Iran, Fars, Shirez	T91063	Av 2
	Pakistan, Parachinar	T91043	Av 3
	Kazakhstan, Kustonoy	T91110	Av 4
	China, Altsi	T91102	Av 5
	China, Yining	T91093	Av 6
	Kazakhstan, Dmitrievka	T91057	Av 7
	Uzbekistan, Tashkent	T90076	Av 8

buffer (10 m*M* Tris HCl, pH 8.0; 1 m*M* EDTA; 1% Nonidet and 100 µg/ml proteinase K), diluted with 48 µl of sterile distilled water, heated to 65°C for 20 min followed by 95°C for 3 min, and centrifuged. Two microliters of RNAse (10 mg/ml) were added to the supernatant (about 80 µl) and kept at room temperature for 15 min. Ten microliters were used as template DNA for PCR or stored at –20°C for later use.

C. PCR TECHNIQUE

Though standard PCR conditions[20] will yield RAPD bands, the results may vary among repetitions of the experiments. Therefore, we have determined PCR amplification conditions for our samples that yield reproducible results. Twenty 10-bp primers of Kit A (OPA-01 to OPA-20) were purchased from Operon Technologies, Inc., Almeida, CA (Table 2), and all other PCR components were purchased from Perkin Elmer Cetus Corporation, Norwalk, CT. A 50-µl reaction mixture in a sterile 500-µl Eppendorf tube contained PCR buffer (10 m*M* Tris-HCl, pH 8.3; 50 m*M* KCl), 3.0 m*M* $MgCl_2$, 200 µ*M* of each of the four deoxynucleoside triphosphate, 40 to 60 pmol of single primer (10-mer), 10 µl genomic DNA template, and 2.5 U of *Taq* polymerase. The mixture was overlaid with 50 µl of mineral oil (Sigma Chemical Co.).

The negative control reaction contained water in place of the genomic DNA. The PCR was carried out in Precision Scientific GTC-2. Each PCR cycle consisted of 92°C for 30 s, 35°C for 1 min, 5 min ramp to 72°C, and 72°C for 2 min. This cycle was repeated 45 times, followed by 72°C for 7 min and cooling down to and maintaining

TABLE 2
The Primers (Kit A of Operon Technologies, Inc.) Used in RAPD Experiments

Code	Sequence (5′-3′)	Code	Sequence (5′-3′)
OPA-01	CAGGCCCTTC	OPA-11	CAATCGCCGT
OPA-02	TGCCGAGCTG	OPA-12	TCGGCGATAG
OPA-03	AGTCAGCCAC	OPA-13	CAGCACCCAC
OPA-04	AATCGGGCTG	OPA-14	TCTGTGCTGG
OPA-05	AGGGGTCTTG	OPA-15	TTCCGAACCC
OPA-06	GGTCCCTGAC	OPA-16	AGCCAGCGAA
OPA-07	GAAACGGGTG	OPA-17	GACCGCTTGT
OPA-08	GTGACGTAGG	OPA-18	AGGTGACCGT
OPA-09	GGGTAACGCC	OPA-19	CAAACGTCGG
OPA-10	GTGATCGCAG	OPA-20	GTTGCGATCC

4°C until used for electrophoresis. Ten microliters of each amplified product were used for electrophoresis in 1.6% agarose using tris borate EDTA (TBE) buffer.[21]

Adjustments in PCR conditions were made to increase or decrease the number of different amplification products (bands) according to suggestions by the supplier of primers (Operon Technologies, Inc.). The less stringent conditions described above were used to generate a large number of bands to determine the level of intraspecies variability. However, in order to decrease the number of bands to facilitate species diagnostics, lower concentrations of $MgCl_2$ (ranging from 3.0 to 0.5 m*M*) were also tested.

D. DATA ANALYSIS

With 20 or more amplified fragments produced by the majority of primers used, it would be too cumbersome to characterize the RAPD-PCR pattern of individual insects in detail. Therefore, for the sake of simplicity, we have assigned the RAPD fragments into four zones based on their migration (approximate size in basepairs) in the gel relative to the molecular size markers: e.g., zone A — fragments in the size range up to 500 bp; zone B — >500 to 1000 bp; zone C — >1000 to 1500 bp; and zone D — fragments larger than 1500 bp. For preparation of the taxonomic key, only reproducible bands were used. Three molecular size markers, e.g., 100-bp ladder, pBR322 DNA-HaeIII, and oX174RF-HaeIII, were used to estimate the sizes of amplified fragments. Fragments of relatively high, medium, and low intensities were scored as H, M, and F respectively, to facilitate their comparison of reproducibility in different PCR experiments. We used the statistic $F = 2N_{xy}/(N_x + N_y)$, commonly used in RFLP data analysis to estimate genetic similarities among the four species. N_x and N_y represent the number of RAPD fragments in individual insects x and y, respectively; N_{xy} is the number of fragments shared by x and y individuals. We used all bands generated by 20 primers in each species (different strains) for this statistics. A zero value for F indicates no shared fragments, whereas a value of 1.00 indicates identical fragments in x and y. The F values are expressed in percentage in Table 7.

III. RESULTS

A. SPECIES-SPECIFIC RAPD FRAGMENTS

All primers produced multiple fragments (range 8 to 26 for 20 primers) with a wide range in sizes (50 to >2500 bp) as shown in Table 3. Although RAPD patterns produced by most primers could be used to distinguish the four species, three of the primers (OPA-03, OPA-12, and OPA-20) produced some fragments unique to all strains of individual species (Table 1). Therefore, species-specific RAPD patterns generated by three primers are presented in detail.

PCR with OPA-12 and OPA-20 produced some prominent and reproducible fragments in each species, including some which were unique to each species (Figures 1 to 3). These included (size estimate in basepairs) 540 in *Aphelinus asychis*, 580 in *A.* sp. nr *varipes*, 780 to 800 in *Diaeretiella rapae*, and 850 in *Aphidius matricariae* with OPA-12. The diagnostic fragments with OPA-20 included 310 in *Aphelinus asychis*; 930 in *A.* sp. nr *varipes*; combination of two fragments, 1050 and 1350 in *Aphidius matricariae*; and 1200 and 1350 in *D. rapae*. The 600 fragment of OPA-19 stands out in *D. rapae* (see Figure 1D and Table 5).

Diagnostic RAPD-PCR patterns were produced with OPA-03 in certain regions relative to marker bands of the 100-bp molecular marker (Figure 4 and Table 4). For example, within the region covering >500 to 600 bp size in zone B, species-specific patterns were comprised of only one fragment, 530 in *Aphelinus asychis*; a wide fragment, 550 (duplet, 550 + 575 in some individuals) in *A.* sp. nr *varipes*; 520 and 580 in *Aphidius matricariae*; and 510 and 550 in *D. rapae*. Similarly, diagnostic patterns were observed in 200 to 400 bp and >400 to 500 bp regions of zone A and >600 to 700 bp region of zone B.

B. DICHOTOMOUS TAXONOMIC KEYS

Based on the presence of unique fragments and patterns produced with OPA-12 and OPA-20, either of the two taxonomic keys (Tables 5 and 6) distinguished the four species. In addition, the schematic drawing of a part of zone A (Figure 4) or patterns produced by OPA-03 (Table 4) or both can be used for identification of species. Lowering the concentration of $MgCl_2$ (from 3.0 to 0.5 m*M*) in PCR experiments resulted in a decrease of the overall number of RAPD fragments in each individual, making the identification of diagnostic fragments easier for taxonomic purpose. But our standard PCR conditions (3.0 m*M* $MgCl_2$) were found most suitable for detecting polymorphism aimed at providing markers for distinguishing geographical strains.

C. POLYMORPHISM

Two types of polymorphisms were observed in RAPDs generated by the primers OPA-01 to OPA-20. The first consisted of the presence or absence of high intensity fragment(s). The second consisted of marked variation in degrees of staining intensity, ranging from the very intense in some to very faint in other individuals of the same species. Both of these classes of fragments were not used for taxonomic keys. Some easily identifiable polymorphic (present in some individuals, absent in others)

TABLE 3
Number (Range and Mean) and Range of Sizes (bp) of Amplification Products Produced by RAPD-PCR Experiments

Species	No Bands Range	Mean	Size range	No Bands Range	Mean	Size range	No Bands Range	Mean	Size range	No Bands Range	Mean	Size range
	Primer OPA-01			Primer OPA-06			Primer OPA-11			Primer OPA-16		
Am	14–19	18.3	140–2500	19–24	22.2	130–2100	16–17	16.3	160–1700	15	15.0	205–1150
Dr	17–22	19.0	170–2100	18–21	19.5	120–2400	19–21	20.0	150–1700	19	19.0	140–1700
Aa	10–19	14.7	140–1300	16–19	18.0	125–2100	19–20	19.3	150–2100	–		
Av	14–23	18.0	140–1250	19–24	21.8	125–1700	17–19	18.3	100–2100	–		
	Primer OPA-02			Primer OPA-07			Primer OPA-12			Primer OPA-17		
Am	23–24	23.3	180–>2300	15–17	16.3	125–2400	18–22	20.1	160–2000	11–14	12.3	175–1350
Dr	19–21	20.0	160–>2300	14–17	15.0	140–2500	20–27	23.0	100–2500	11–17	14.0	120–1700
Aa	23–26	24.0	110–>2100	19–21	20.0	125–2100	21–23	22.3	140–2500	18	18.0	240–1900
Av	21–24	22.0	50–1800	14–20	17.3	140–1700	19–23	21.0	120–2200	13–17	15.0	260–1100
	Primer OPA-03			Primer OPA-08			Primer OPA-13			Primer OPA-18		
Am	16–24	20.5	220–2200	17–23	20.3	160–>2300	13–14	13.7	200–1350	15–21	18.7	110–2000
Dr	19–24	20.6	150–2200	21–23	22.0	120–>2300	12–17	15.0	150–1800	18–20	19.0	110–2000
Aa	18–27	22.1	100–2100	20–23	21.3	120– 2100	16–17	16.7	190–1300	19–21	20.3	150–2100
Av	19–24	21.1	105–2100	20–24	22.0	140– 2200	15–16	15.7	150–1050	19–20	19.5	160–2000

	Primer OPA-04			Primer OPA-09			Primer OPA-14			Primer OPA-19		
Am	9–14	10.8	120–1700	4	4	275–510	14–17	14.7	280–1350	26–29	27.7	100–1800
Dr	13–14	13.5	160–1500	8–10	9	140–800	13–18	15.3	220–1900	18–21	19.4	90–2200
Aa	15	15	170– 800	12–17	14	210–2000	14–16	14.7	260–1200	24–26	24.8	130–1500
Av	7–16	10.8	150–1025	13	13	110–1050	15–16	15.7	220–1350	25	25.0	~90–1400
	Primer OPA-05			**Primer OPA-10**			**Primer OPA-15**			**Primer OPA-20**		
Am	19–25	21.7	175–2200	16–17	16.3	180–1400	13–16	14.0	180–1350	17–20	19.1	170–3000
Dr	18–29	23.4	180–2100	15–20	18.0	180–1400	17–19	18.0	130–2200	15–20	18.8	170–2500
Aa	18–20	18.7	130–2300	17–18	17.3	150–1250	15–20	18.3	210–1800	18–23	21.3	120–2500
Av	17–24	20.7	160–2300	12–18	14.7	160–1600	14–17	15.3	160–1800	22–26	23.4	120–2500

Note: Am, *Aphidius matricariae*; Dr,*Diaeretiella rapae*; Aa, *Aphelinus asychis*; and Au, *Aphelinus* sp. *nr varipes*.

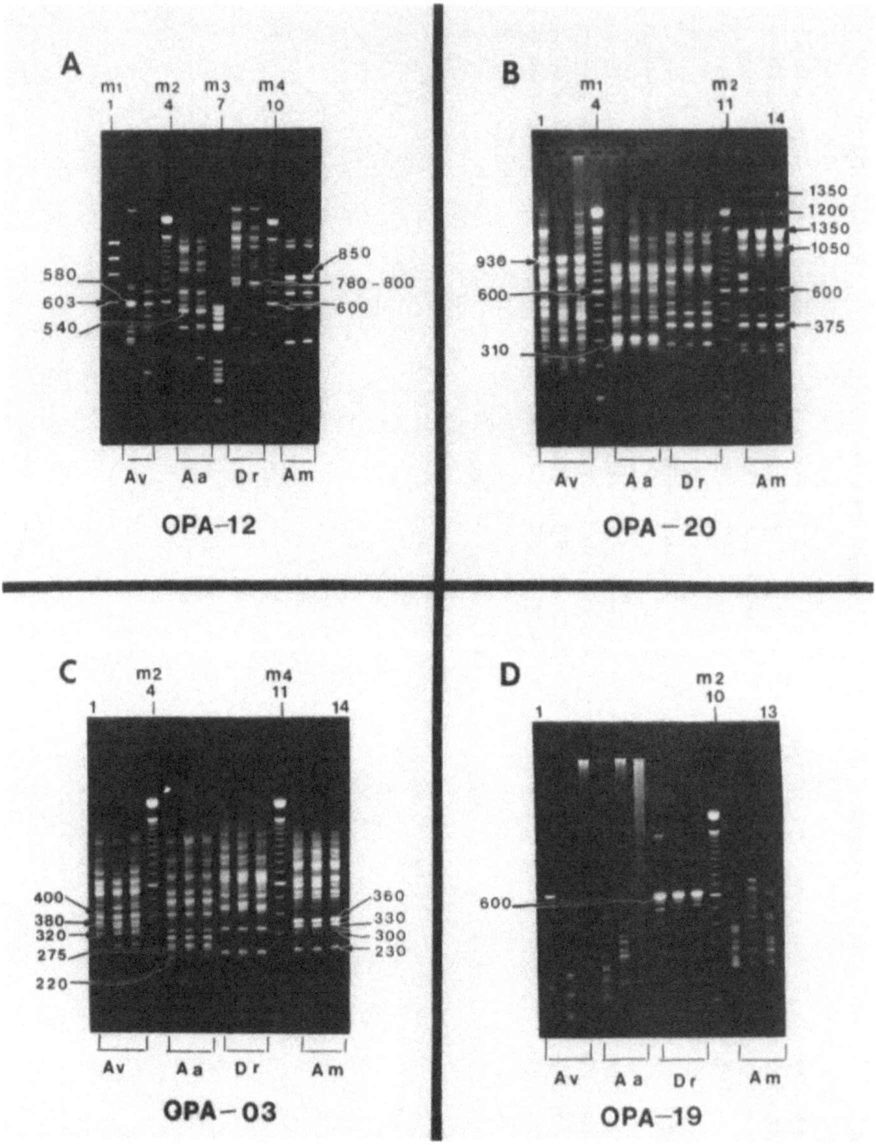

FIGURE 1. Comparison of RAPD-PCR banding patterns of four species of parasitic Hymenoptera using four primers:OPA-03, OPA-12, OPA-19, and OPA-20. Molecular size markers include m1 (øX174 DNA-HaeIII), m2 (100-bp ladder), and m3 (pBR322 DNA-HaeIII). Am: *Aphidius matricariae*, Dr: *Diaeretiella rapae*, Aa: *Aphelinus asychis*, and Av: *A.* sp. nr *varipes*. Note the most distinct 600-bp fragment of *D. rapae* in lanes 7 to 9 (part D).

fragments generated by the primer OPA-12 include 300, 480, 560, 590, 950, and 1350 in *Aphidius matricariae*; 280, 375, 425, 820, and 1600 in *D. rapae*; 280, 400, 425, 550, 680, 1150, 1200, and 1400 in *Aphelinus asychis*; and 280, 330, 550, 975,

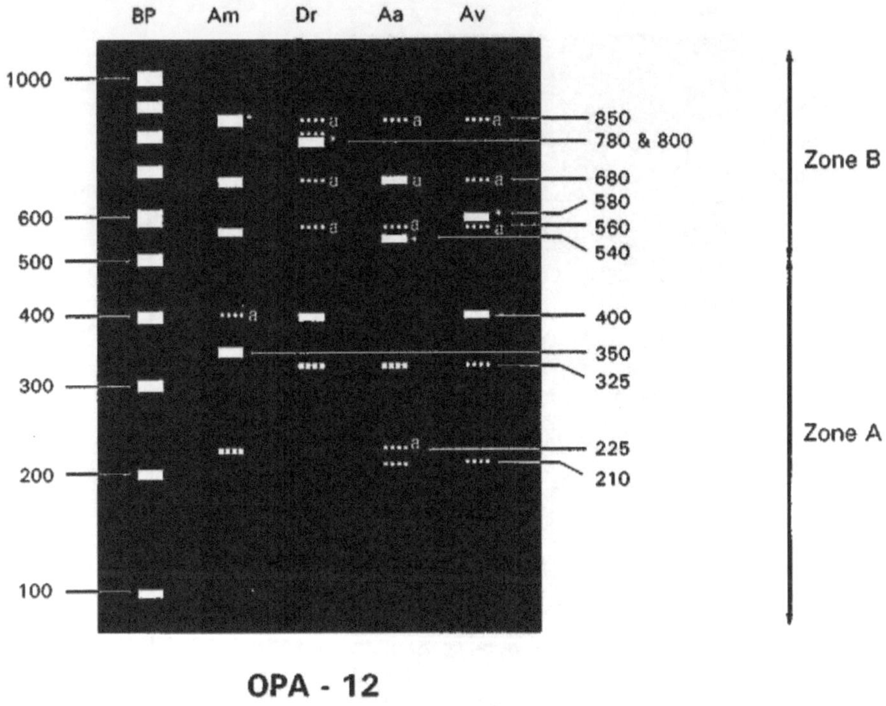

FIGURE 2. A schematic drawing showing the comparison of some reproducible RAPD fragments in four species using the primer OPA-12 (complete patterns in Figure 1A). The prominent fragments indicated by the asterisk are useful taxonomic markers. Those marked with "a" show variable intensities. A part of the molecular size marker, 100-bp ladder (BP), used for estimation of sizes is included. The species codes are the same as in Figure 1.

and 1200 in *A.* sp. nr *varipes*. Similarly, polymorphic fragments generated by the primer OPA-20 include 550, 590, 700, 750, 875, and ≤1000 in *Aphidius matricariae*; 430, 460, 500, 575, 650, 780 (780 to 820), 980 (980 to 1000), and 1600 in *D. rapae*; 430, 600, 1100 (1100 to 1150), and 1250 in *Aphelinus asychis*; and 260, 275, 400, 600, 630, 700, 1400, and 1600 in *A.* sp. nr *varipes*.

Analysis of RAPDs (20 primers) from individual insects revealed 89% monomorphic and 11% polymorphic fragments in *Aphidius matricariae* and 72% monomorphic and 28% polymorphic in *D. rapae*. In each of the two species, *Aphelinus asychis* and *A.* sp. nr *varipes*, 78% of RAPD fragments were monomorphic and 22% were polymorphic.

D. GENETIC SIMILARITIES

Table 7 shows the percent similarities within and between the four species of parasitic Hymenoptera based on RAPD patterns with 20 primers. Two species of the same genus, *Aphelinus*, shared a higher percent of RAPD fragments (F: 46%), as compared to species of different genera (range 26 to 32%). Of the four species, *Aphidius matricariae* showed less variability (F: 89%) than others (F: 72 to 78%).

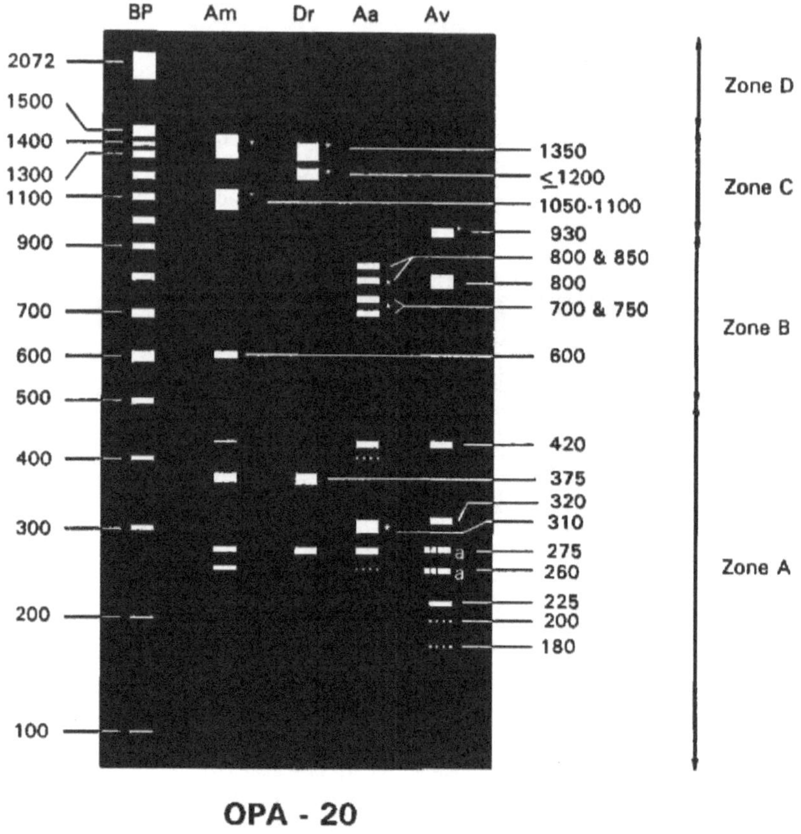

FIGURE 3. A schematic drawing showing the comparison of some reproducible RAPD fragments in four species using the primer OPA-20 (complete patterns in Figure 1B). The prominent fragments indicated with the asterisk are useful taxonomic markers. Those marked with "a" show variable intensities. A part of the molecular size marker, 100-bp ladder (BP), used for estimation of sizes is included. The species codes are the same as in Figure 1.

IV. DISCUSSION

The principal objective of this study was to assess the potential application of RAPD-PCR in the study of DNA poymorphisms and in taxonomic identification of four species of parasitic Hymenoptera. Our results show that Hymenoptera, which generally reveal low levels of allozyme polymorphisms, show numerous RAPDs. We are extending RAPD studies to the analyses of some sequences of rRNA (ribosomal RNA) and tRNA (transfer RNA) for distinguishing biotypes of species of parasites and predators of the RWA.

The RAPD-PCR method is gaining wide application in studies on taxonomy and population structure (e.g., sibling species, discrete subpopulations, subspecies, host races, biotypes, ecotypes, strains, etc.). It has both advantages and disadvantages over allozyme and restriction fragment length polymorphism (RFLP)-Southern blotting techniques. The advantages include the following.

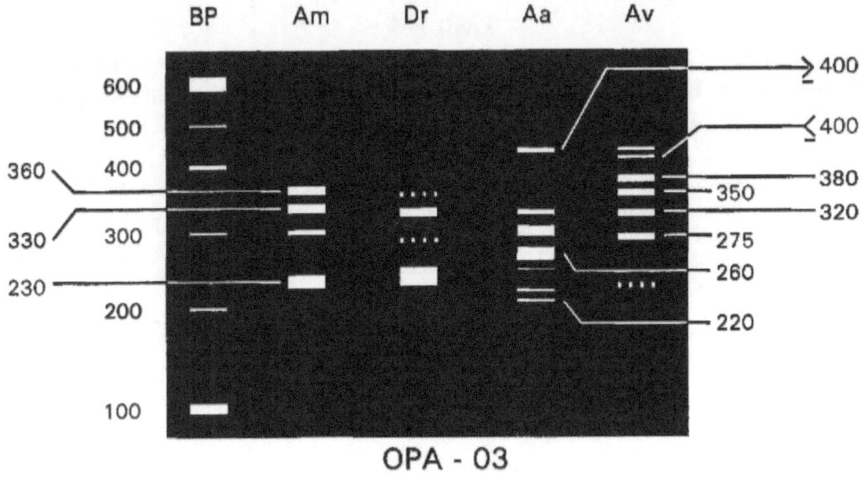

FIGURE 4. A schematic drawing showing the comparison RAPD patterns of four species of parasitic Hymenoptera using the primer OPA-03 (complete patterns in Figure 1C). A part of the molecular size marker, 100-bp ladder (BP), used for estimation of sizes is included. The species codes are the same as in Figure 1.

TABLE 4
Some Easily Identifiable and Reproducible Differences in RAPD Bands Produced by Amplification of DNA of Four Species[a] by OPA-03 Primer

	Zone A regions		Zone B regions	
Species	200–400 bp	>400–500 bp	>500–600 bp	>600–700 bp
Am	230, 300, 330, 360	≤500	520, 580	625D
Dr	230, 320	430, 450 (or 450 and 460)	510, 550	630, 660, 700
Aa	220, 260, 280, 320	≥400, 430(v), 480	530	660
Av	230, 275, 320, 350, 380, ≤400	410, 450 (or 460 or 450 and 460)	550 (550 + 575)	≥625 + 650

Note: The high intensity bands are underlined. V, variable expression (faint to medium). The "+" sign indicates that the bands can sometime appear as a single wide band. D indicates that the band is made up of two fine bands.

[a] For species code see Table 1.

1. RAPD analysis is simpler, faster, and demands less technical expertise than RFLP analysis.
2. The amount of DNA required for RAPD-PCR is 1% or less than that required for RFLP analysis.
3. The same single short primers usually can be used for analysis of widely different taxonomic groups, whereas use of RFLP probes is restricted by the degree of homology to the DNA of the target taxon.
4. Prior sequence data of the target DNA for generation of primers is not necessary for amplification.

TABLE 5
Dichotomous Taxonomic Key for Identification of Four Species of Parasitic Hymenoptera, Based on RAPD Fragments (Approximate Sizes in Basepairs) Using the Primer OPA-12 (Figures 1A and 2)

1. Fragment 580 more distinct than any other fragment .... *Aphelinus* sp. nr *varipes*
   Fragment 580 either faint or absent 2
2. Fragment 540 is distinct .... *A. asychis*
   Fragment 540 either faint or absent 3
3. Fragment 850 distinct; duplet 780 + 800 absent .... *Aphidius matricariae*
   Fragment 850 faint or absent but duplet 780 + 800 present .... *Diaeretiella rapae*

Note: RAPD-PCR analysis of individuals with both OPA-12 and OPA-19 would further strengthen identifications of *D. rapae*. The fragment, 600 (OPA-19), of *D. rapae* is most distinct and is about three times the width of the 600 of the 100-bp molecular marker (Figure 1D).

TABLE 6
Dichotomous Taxonomic Key for Identification of Four Species of Parasitic Hymenoptera, Based on RAPD Fragments (Approximate Sizes in Basepairs) Using the Primer OPA-20 (Figues 1B and 3)

1. Fragment 310 is more distinct than any other fragment .... *Aphelinus asychis*
   Fragment 310 faint or absent 2
2. Fragment 930 distinct .... *A.* sp. nr *varipes*
   Fragment 930 faint or absent 3
3. Fragments 1050 and 1350 (twice as wide as 1050) more distinct than other fragments .... *Aphidius matricariae*
   Fragment 1050 faint or absent, but fragments 1200 and 1350 more distinct than those in *A. matricariae* .... *Diaeretiella rapae*

TABLE 7
Percent Similarities (%F) in RAPD Fragments Within and Between Four Species of Parasitic Hymenoptera

		1	2	3	4
1.	*Aphidius matricariae*	100	33	27	26
2.	*Diaeretiella rapae*		100	27	32
3.	*Aphelinus asychis*			100	46
4.	*A.* sp. nr *varipes*				100

5. The availability of a large number of random primers (800 10-bp primers are available from Operon Technologies, Inc.) makes RAPD-PCR a powerful tool for detecting numerous polymorphisms, which are useful for population genetics and systematics studies.
6. Relatively crude DNA can be used for amplification; even the DNA of preserved specimens can be amplified by PCR.

In addition, discrete data (presence or absence of DNA fragments) are suitable for statistical analysis of genetic relatedness (similarity index, percent match analysis, and genetic distance matrices including Nei's unbiased genetic distance) among taxa. The nearest-neighbor analysis of frequency of fragments can be used to select a group of fragments for discrimination of subpopulations and possibly populations.

There are some disadvantages in using the RAPD-PCR approach for population genetics, genetic mapping, and taxonomic studies. First, a large number of DNA fragments are generated by most random primers. Generally, a high level of polymorphism is also detected. This can make the identification of taxonomic markers a very difficult task. The number of fragments for each PCR depends upon the primer sequence, genomic sequence, degree of mismatch between the very short primer (10 bp) and template DNA, and the genome size. Black et al.[13] reported an enormous amount of genetic variation in four aphid species. William et al.[11] documented that about 50% of the 11 RAPD fragments examined were from repetitive sequences in *Glycine soja* and *G. max*.

First, the generation of a large number of RAPD bands can make identification of taxonomic markers cumbersome. Therefore, it is desirable to reduce the total number of bands so that key bands can be easily recognized. The number of bands can be reduced by changing the PCR conditions, e.g., increasing the annealing temperature, and reducing the concentration of $MgCl_2$. By lowering the concentration of $MgCl_2$ from 3.0 to 0.05 m*M*, we were able to reduce considerably the number of fragments with OPA-12 and OPA-20, while still retaining those used in the taxonomic key.

Second, many fragments (especially those arising from mispairing of a primer with the genomic DNA) may not be reproducible between different laboratories because amplifications are more sensitive to slight changes in temperature cycle, particularly the annealing temperature, and differences in temperature of wells in the same and between tempcyclers. Practical considerations for conducting the RAPD analysis of genomes have been discussed by Arnheim.[22]

Third, in contrast to RFLP (on autoradiographs), which are usually codominant, most of the polymorphisms detected by RAPD-PCR are inherited as dominant markers.[11] Therefore, homozygotes and heterozygotes cannot be differentiated. In RAPD-PCR, both genotypes will show one band because the size difference is more likely to appear as the presence or absence of a band. Two copies of amplifiable genomic region are indistinguishable from one copy (of that region) in heterozygotes.

Fourth, RFLP bands of the same size represent closely related sequences, but the RAPD bands (on electrophoresis gel) of the same size may not actually have same or similar sequence. Comigrating RAPD bands may not be allelic.

Finally, the RAPD-PCR technique is prone to artifacts resulting from contamination of the PCR mixture by small amounts of foreign DNA (from cross-contamination from other individual insects, microbes and fungi from infected tissues, ecto- and endoparasites, etc).

In contrast to the successful application of RAPD-PCR approach in taxonomic identification of species, host races,[13] geographical populations, and subspecies,[18] Raich et al.[23] were unable to detect species-specific RAPD bands in some North American *Culicoides* species. They reported a high level of polymorphism in all species examined. It was assumed that the testing of additional primers might lead to detection of bands unique to different species of *Culicoides*.

In summary, although RAPD-PCR may not provide directly and as easily accessible information on genetic relationships as RFLP data, it is superior to RFLPs in uncovering cryptic genetic variability in organisms for distinguishing difficult taxa.

ACKNOWLEDGMENTS

This work was supported, in part, by funds from APHIS of the U.S. Department of Agriculture. We thank Dr. David Prokrym for useful suggestions during implementation of this research and Leeda Wood and Debby Nelson for shipping specimens for PCR analysis. Thanks to Judy Eaton and Mark Matuska for technical help and to Dianne Evanson, Bonnie Muhl, and Genny Olstad for assistance in preparation of illustrations. Mention of a trademark or proprietary product does not constitute a guarantee or warranty of the product by the U.S. Department of Agriculture and does not imply its approval to the exclusion of other products that may also be suitable.

REFERENCES

1. **Hillis, D. M. and Moritz, C., Eds.,** *Molecular Systematics*, Sinauer Associates, Inc., Sunderland, MA, 1990.
2. **Murphy, R. W., Sites, J. W., Jr, Buth, D. G., and Haufler, C. H.,** Protein I: isozyme electrophoresis, in *Molecular Systematics*, Hillis, D. M. and Moritz, G., Eds., Sinauer Associates, Inc., Sunderland, MA, 1990, 82.
3. **Narang, S. K. and Seawright, J. A.,** Electrophoretic method for recognition of sibling species of anopheline mosquitos. A practical approach, *Fla. Entomol.*, 71, 303, 1988.
4. **Narang, S. K.,** Molecular approaches for the identification of population structures of economically important insects, in *The Eukaryotic Chromosome: Structural and Functional Aspects*, Vol. 2, Sobti, R. C. and Obe, G., Eds., Springers-Norosa Publication Co., New Delhi, 1994.
5. **Graur, D.,** Gene diversity in hymenoptera, *Evolution*, 39, 190, 1985.
6. **Bloch, W.,** A biochemical perspective of the polymerase chain reaction, *Biochemistry*, 30, 2735, 1991.
7. **Demeke, T., Adams, R. P., and Chibbar, R.,** Potential taxonomic use of random amplified polymorphic DNA (RAPD): a case study in *Brassica, Theor. Appl. Genet.*, 84, 990, 1992.
8. **Caetano-Anolles, G., Bassam, B. J., and Gresshoff, P. M.,** High resolution DNA amplification fingerprinting using very short arbitrary oligonucleotide primers, *Bio-Technology*, 9, 553, 1991.
9. **Welsh, J. and McClelland, M.,** Fingerprinting genomes using PCR with arbitrary primers, *Nucl. Acids Res.*, 18, 7213, 1990.
10. **Welsh, J., Petersen, C., and McClelland, M.,** Polymorphisms generated by arbitrarily primed PCR in the mouse: application to strain identification and genetic mapping, *Nucl. Acids Res.*, 19, 303, 1991.
11. **Williams, J. G. K., Kubelik, A. R., Livak, K. J., Rafalski, J. A., and Tingey, S. V.,** DNA polymorphisms amplified by arbitrary primers are useful as genetic markers, *Nucl. Acids Res.*, 18, 6531, 1990.
12. **Chapco, W., Ashton, N. W., Martel, R. K. B., and Antonishyn, N.,** A feasibility study of use of random amplified polymorphic DNA in the population genetics and systematics of grasshoppers, *Genome*, 15, 569, 1992.

13. **Black, W. C., IV, DuTeau, N. M., Puterka, G. J., Nechols, J. R., and Pettorini, J. M.,** Use of the random amplified polymorphic DNA polymerase reaction to detect DNA polymorphisms in aphids, *Bull. Entomol. Res.*, 82, 151, 1992.
14. **Gawel, N. J. and Bartlett, A. C.,** Characterization of differences between whitefly species using RAPD-PCR, *Insect Mol. Biol.*, 2, 33, 1993.
15. **Heckel, D. G.,** Classical and molecular approaches to genome mapping in Lepidoptera, Paper presented at The 37th Annual Meeting of the Entomological Society of America, New Orleans, LA, Dec 2–6, 1990.
16. **Heckel, D. G.,** Comparative genetic linkage mapping in insects, *Annu. Rev. Ent.*, 38, 381, 1993.
17. **Kambhampati, S., Black, W. C., and Rai, K. S.,** Randomly amplified polymorphic DNA of mosquito species and populations (Diptera: Culicidae). Technique, statistical analysis and applications, *J. Med. Entomol.*, 29, 939, 1992.
18. **Ballinger-Crabtree, M. E., Black, W. C., IV, and Miller, B. R.,** Use of genetic polymorphisms detected by the random-amplified polymorphic DNA polymerase chain reaction (RAPD-PCR) for differentiation and identification of *Aedes aegypti* subspecies and populations, *Am. J. Trop. Med. Hyg.*, 47, 893, 1992.
19. **Simon, C., Franke, A., and Martin, A.,** The polymerase chain reaction: DNA extraction and amplification, in *Molecular Taxonomy*, Vol. H57 Hewett, G. M., Johnston, A. W. B., and Young, J. P. W., Eds., NATO Advanced Studies Institute, Springer-Verlag, Berlin, 1991, 329.
20. **Innis, M. A. and Gelfand, D. H.,** Optimization of PCRs, in *PCR Protocols: A Guide to Methods and Applications*, Innis, M. A., Gelfand, D. H., Sninsky, J. J., and White, T. J., Eds., Academic Press, New York, 1990.
21. **Sambrook, J., Fritsch, E. F., and Maniatis, T.,** *Molecular Cloning, A Laboratory Manual,* Vols. 1–3, Cold Spring Harbor Laboratory Press, Cold Spring Harbor, NY, 1989.
22. **Arnheim, N.,** Polymerase chain reaction strategy, *Annu. Rev. Biochem.*, 61, 131, 1992.
23. **Raich, T. J., Archer, J. L., Robertson, M. A., Tabachnick, W. J., and Beaty, B. J.,** Polymerase chain reaction approaches to *Culicoides* (Diptera: Ceratopogonidae) identification, *J. Med. Entomol.*, 30, 228, 1993.

Chapter 4

GENETICS OF PARASITIC HYMENOPTERA

Robin E. Owen

TABLE OF CONTENTS

0-8493-2607-9/94/$0.00+$.50

I. INTRODUCTION

Parasitic or parasitoid Hymenoptera are widely used for biological control; indeed, over 360 species representing 18 families have been used in control programs.[1] Although many details of their biology obviously differ, the vast majority of hymenopteran parasitoids share one feature in common — their genetic system of haplodiploidy — in which they also differ from most other insects. In this chapter, I discuss various features of parasitoid genetics which arise as a consequence of haplodiploidy and are relevant (although not exclusively so) to biological control. The topics to be covered are (1) genetic variation; (2) sex determination and diploid males; (3) sex ratios and mating systems; (4) geographical variation, host-parasite coevolution, and sexual dimorphism; and (5) extrachromosomal inheritance.

II. HAPLODIPLOIDY

In most Hymenoptera, females arise from fertilized eggs, while males are produced by arrhenotoky (parthenogenesis) from unfertilized eggs and are therefore haploid. However, *diploid* males can also be produced as a result of the underlying genetic method of sex determination which occurs in some species. This will be discussed in Section IV. Thelytoky, the production of diploid females by parthenogenesis, occurs in a minority of species and will not be discussed here.

III. GENETIC VARIATION

A. SINGLE GENES

Visible polymorphisms are well known in a few species. In *Bracon hebetor* (Braconidae, formerly *Habrobracon juglandis*), probably the best known parasitoid genetically, about 100 different naturally occurring visible mutations have been described.[2] These include eye and body color mutations and morphological mutations involving wing, head, and antennae.[3] Ionizing radiation and chemical mutagens have also been used to generate visible and lethal (dominant and recessive) mutations.[2] Some genetic work has also been done on *Habrobracon serinope* (for a comprehensive review of Braconid genetics see Reference 2).

Numerous spontaneous and induced mutations are known in the Chalcidoid *Nasonia vitripennis* (formerly *Mormoniella*).[3] Many of these are eye-color mutants, and linkage maps of 47 mutants are given by Saul, Saul, and Becker.[4]

These species notwithstanding, classical genetic variation of this type is rare in parasitoids, as in most other organisms. However, the use of electrophoresis has enabled variation at the biochemical level to be examined and estimates of the level of genetic variation (at least at structural gene loci) to be made. Electrophoretic variation has been found in over 20 species of parasitoid, and in most of these the average heterozygosity (H_{exp}) has been estimated (Table 1). The heterozygosity expected on the basis of Hardy-Weinberg proportions is calculated at each locus and

TABLE 1
Allozyme Variation in Parasitoid Hymenoptera

Species	N[a]	n[b]	P[c]	H_{exp}	Ref.
Opius juglandis	6	13	0.23	0.043	78
Ageniaspis fusciollis	1	25	0.04	0.019	79
Diadegma armillata	1	19	0.16	0.048	79
Itoplectis maculator	1	20	0.15	0.056	79
Pimpla turionellae	1	29	0.14	0.038	79
Triclistus yponomeutae	1	20	0.10	0.020	79
Trieces tricarinatus	1	14	0	0.000	79
Tetrastichus evonymellae	1	20	0.15	0.055	79
Muscidifurax raptor	5	10	0.40	0.066	80
M. zaraptor	3	10	0.40	0.067	80
Spalangia cameroni	4	15	0.13	0.007	80
S. endius	3	15	0	0.000	80
S. nigroaenea	3	15	0.07	0.027	80
Mesochorus nigripes a[d]	1	12	0.7	0.187	81
M. nigripes t[d]	1	12	0.4	0.103	81
Aphidius ervi	9	16	0.44	0.094	82
A. pisivorus	1	16	0.19	0.074	82
A. eadyi	3	16	0.19	0.016	82
A. smithi	2	16	0.12	0.010	82
A. "KR"	1	16	0.06	0.008	82
A. picipes	1	16	0	0.000	82
Diadromus pulchellus	1	22	0.27	*[e]	83
Microplitis croceipes	1	37	0.22	0.212	84

$\bar{H} = 0.052 \pm 0.057$ (S.D.)

[a] Number of populations studied.
[b] Number of loci.
[c] Proportion of loci polymorphic at 5% level.
[d] *a* = Arrhenotokous form, *t* = thelytokous form.
[e] * This is a laboratory population; therefore, heterozygosity is not calculated.

then averaged across all loci, with monomorphic ones included. Heterozygosity in parasitoids is quite low, though comparable to that found in the Hymenoptera in general.[5,6] In the most recent survey of the literature, Crespi[7] calculated an $\bar{H}_{exp}$ of 0.048 ± 0.010 (S.E.) for 58 species of Hymenoptera. Hymenoptera have significantly lower levels of allozyme variation than comparable diplodiploid insects ($\bar{H}_{exp} = 0.120 \pm 0.006$)[5], and this is predicted as a consequence of haplodiploidy.

Formally, haploldiploidy is equivalent to sex-linked inheritance. It is well known that equilibrium frequencies at loci undergoing selection and/or random drift and mutation will be lower at X-linked loci as compared to autosomal loci.[8]

Selection acts to reduce genetic variability in two ways. First, deleterious alleles are removed more quickly from populations because they are exposed to selection in each generation of the haploid males. Second, the condition for a balanced polymorphism is more stringent at an X-linked locus than at an autosomal locus. Thus, with genotypes and viabilities,

	Female			Male	
Genotype	A_1A_1	A_1A_2	A_2A_2	A_1	A_2
Viability	w_{11}	w_{12}	w_{22}	v_1	v_2

the condition for a balanced polymorphism is

$$\tfrac{1}{2}w_{12}(v_1+v_2)>w_{11}v_1,\ w_{22}v_2$$

Because the viabilities of three diploid and two haploid genotypes are involved, the probability of a polymorphism occurring is significantly lower than that at an autosomal locus.[9-11] This is true even when the possibility of selection acting in opposite directions in the two sexes at autosomal loci is taken into account. Similarly, the conditions for stable equilibrium with selection at two loci are more stringent at X-linked loci than at autosomal loci.[12]

Random genetic drift is expected to be more rapid at X-linked and haplodiploid loci due to a smaller effective population size than at comparable diploid loci. In a population consisting of M males and F females, the sex-linked effective population size N_{ex} = 9MF/2(2M + F), and for autosomal loci N_{ea} = 4MF/(M + F). N_{ex} will be less than N_{ea} unless F >7M.[13] Nagylaki[14] considered the joint effects of drift and mutation to derive the expected frequency of heterozygote females assuming an infinite allele model (each new mutation produces a new allele),

$$E(H)=4N_{ex}\mu/(1+4N_{ex}\mu)$$

where μ is the mutation rate.

This expression is equivalent to that for an autosomal locus[15] except for the sex-linked effective population size. Consequently, with mutation and drift, the expected frequency of heterozygotes is lower at X-linked and haplodiploid loci than at diploid loci.[8]

Given that both selection and mutation/drift lead to the same predictions concerning genetic variability, it is difficult to determine the relative influence of each. One test of the mutation/drift hypothesis is suggested by Avery[8] who following Nagylaki[14] gives an approximation for the variance in heterozygosity at a single locus,

$$\text{VarH}\cong 8N_{ex}\mu/\left[(1+4N_{ex}\mu)^2(2+4N_{ex}\mu)(3+4N_{ex}\mu)\right]$$

The between-species variation in heterozygosity is then approximated by $\text{Var}(\bar{H}) \simeq \text{Var(H)}/\bar{n}$, where $\bar{n}$ is the harmonic mean of the number of loci studied in each species. An important assumption is that the species have approximately similar effective population sizes.[8] Using the data in Table 1, the observed H is equated to

E(H) to estimate $4N_{ex}\mu$ and thus derive Var(*H*).[8] Interestingly, the predicted standard deviation of 0.003 is much less than the observed standard deviation of 0.057, i.e., the observed variance of heterozygosity in these parasitoids is considerably larger than expected; this could result from widely varying N_es. A closer agreement is obtained when the data from the genus *Aphidius* alone is considered (predicted S.D. = 0.026, observed S.D. = 0.040).

The idea that haplodiploidy per se accounts for low levels of genetic variation is supported by the observation that haplodiploid Thysanoptera have heterozygosities ($\bar{H}_{exp} = 0.058 \pm 0.011$[S.E.]) comparable to Hymenoptera.[7] Nevertheless, as far as the Hymenoptera are concerned, other traits and habits may also have an effect on heterozygosity. Unruh et al.[16] investigated the rate of decay of heterozygosity at eight allozyme loci in population cages of *A. ervi*. Populations were established with 5, 15, 30, and 100 mated females, respectively, and each generation of these were reestablished with the same number of females. Uhruh et al.[16] found that on the average heterozygosity was lost at a rate considerably higher than that predicted by the equation

$$H_t = H_0(1 - 1/2N_{ex})^t$$

where H_t is the heterozygosity at generation t and H_0 is the initial heterozygosity.

Effective population size was also estimated using the above equation and found to be only 48% of that predicted on the basis of the actual numbers of individuals in the population [i.e., from 9MF/2(2M + F)]. Unruh et al.[16] suggest that this was most likely due to unequal participation by individuals in reproduction. Parasitoids, of course, are particularly susceptible to reduced N_es due to cyclic fluctuation in population numbers, a common phenomenon in parasitoid-host population dynamics.[17] This is because N_e also depends on the harmonic mean of population size,[18] which is most influenced by the smallest numbers. These ecological and behavioral factors could account for the large variance in heterozygosities observed in parasitoids (Table 1).

In conclusion, electrophoretic studies show that parasitoids, like other Hymenoptera, have quite low levels of allozyme variation. This can be lost quite rapidly in laboratory populations because of the small effective population size.

B. POLYGENES

Although allozyme variation has been assessed in parasitoids, relatively little is known about the extent of quantitative genetic variation. Haploid males are predicted to have twice the population variance as females (assuming dosage compensation). This appears to be the case in polistine wasps[19] and bumble bees.[20]

Few direct investigations of polygenic inheritance have been done on parasitiods, with one exception being the extensive work by Legner[21,22] on *Muscidifurax raptorellus*. This species exhibits polygenically inherited gregarious oviposition behavior, and Legner has demonstrated through breeding experiments that at least 8, and possibly up to 19, loci are involved. The situation is complicated by extranuclear influences transmitted by the male which cause changes in the oviposition behavior of females.[23] More details of this will be discussed in Section VII.

Recently, Sequeira and Mackauer[24] estimated the extent of quantitative genetic variation for body size and development time in *A. ervi*. Using a half-sib design, they found a heritability of 0.38 (averaged over sire and dam components) for body size, but were unable to reliably estimate the heritability of development time due to small sample size. However, because of the large difference the sire and dam components, they did conclude that the additive genetic variance for this trait was probably close to zero.[24] These results are of considerable interest because body size and development time are important components of life history strategies.[24] Body size is correlated with fecundity in female parasitoids and with mating success in male parasitoids.[25]

Indirect evidence of polygenic variation comes from selection for traits useful for biological control. Wilkes[26] successfully selected for high fecundity in *Dahlbominus fuscipennis*. After 11 generations of family selection, he increased the mean number of offspring per female by 40 to 60%. Allen[27] shifted the host preference of *Horogenes molestae* from its natural host, an oriental fruit moth, to an alternate host, the potato tuberworm *Phthorimaea operculella*. White, DeBach, and Garber[28] used mass selection to increase temperature tolerance in *Aphytis lingnanensis*. However, not all attempts at selection succeeded. van Lenteren[29] reported an unsuccessful attempt to select for increased ovariole number in *Encarsia formosa*. The size of the host on which the parasitoid was reared turned out to be the greatest influence on ovariole number. Finally, quantitative genetic variation for life history traits can be inferred from intraspecific (geographical) variation; some aspects of this will be touched on in Section VI, otherwise see Unruh and Messing[30] for a review.

IV. SEX DETERMINATION AND DIPLOID MALES

A. SEX DETERMINATION

Although, at a phenomenological level, sex is determined by ploidy, a genetic mechanism of sex determination has been demonstrated in some Hymenoptera. First discovered in *B. hebetor*,[31] this system involves a single locus with multiple alleles $x_1, x_2, x_3, \ldots, x_k$. Individuals heterozygous for any two alleles become females, and haploid (or hemizygous) individuals become males, but homozygotes also become males which are, of course, diploid. Although in *B. hebetor* this acts as a single locus, it should really be regarded as a supergene, with the molecular basis of the gene action unknown.[2]

Similar single-locus systems have been found in other Hymenoptera, including the bees *Apis mellifera* and *Melipona quadrifasciata*,[32] the sawflies *Neodiprion nigroscutum*[33] and *Athalia rosae rificornis*,[34] and the ichneumonid *Diadromus pulchellus*.[35] Although the single-locus mechanism has been found in only these few species, it or a multiple-locus extension is assumed to be widespread in Hymenoptera.[36]

In the multilocus model, heterozygosity is required at a minimum of one locus for female determination.[36] The multiple-locus model is predicted to apply to species that regularly undergo inbreeding. Skinner and Werren[37] tested this with such a species,

Nasonia vitripennis, but were unable to produced any diploid males after six generations of mother/son inbreeding. This suggests that alternative mechanisms of sex determination occur in some Hymenoptera. However, diploid males have been detected in a number of hymenopteran taxa, and they do presumably result from homozygosity at one or more sex-determining loci. In addition to the breeding studies quoted above, diploid males have been detected cytogenetically and electrophoretically in five genera of ants,[38] two species of sweat bees,[39,40] the braconids *Microplitis croceipes* and *Bathyplectes curulionis*, and the ichneumonid *Liotryphon caudatus*.[36]

B. DIPLOID MALES IN POPULATIONS

Diploid males result from matings between a female and a male having one sex-determination allele in common; thus, from the mating $x_i x_j \times x_i$, females $x_i x_j$ and diploid males $x_i x_i$ will be produced in a 1:1 ratio. These are termed matched matings by Adams et al.[41] Diploid males are functionally useless and thus represent a reduction in reproductive success of the parents and a type of genetic load on the population. However, because in finite populations there can only be a finite number of alleles, some matched matings will inevitably occur even with panmixis. Obviously, inbreeding, either due to consanguinity or small population size, will lead to an increase in the diploid male production. In this section, I will discuss the frequency of diploid males expected in populations of parasitoids and Hymenoptera in general and some of the problems involved in the estimation and interpretation of this.

Considering a random mating population at equilibrium, Adams et al.[41] have shown that the frequency of matched matings is

$$\theta = 2/K$$

where K is the number of sex alleles. In an infinite population, the equilibrium frequency of the alleles is $p_1 = p_2 = p_3 = \ldots = p_k = 1/K$, i.e., all alleles are expected to be at equal frequency.[42] Clearly, selection should act to maximize the number of alleles; however, in finite populations, the number of alleles will be determined by the balance between the gain of new alleles through mutation and the loss of alleles through drift. Yokoyama and Nei[43] determined the effective number of sex alleles (K) maintained in a population of size N_{ex} and constant mutation rate $\mu = 10^{-5}$:

N_{ex}	100	500	1000	5000
K	5.32	11.73	16.88	39.75

In a finite population, the alleles will not be at equal frequencies, but rather they will be distributed around a mean value. The K value given above is the reciprocal of this mean.[43] Clearly, we would expect the number of sex alleles not to be very high even in natural populations of parasitoids and especially not in introduced and laboratory populations founded from a relatively small number of individuals. In fact, estimates of the number of sex alleles found in natural populations range from 5 to 25.[30]

I will now turn to the frequency of diploid males expected in a population. Diploid males can be viewed from two perspectives: either as the proportion of diploids that are male (Φ) or as the proportion of males that are diploid (ϕ). Progeny arise from two sources: fertilized eggs and unfertilized eggs.

	Fertilized eggs	**Unfertilized eggs**
Diploids	Φ diploid males	$(1 - \Phi)$ females
Males	ϕ diploid males	$(1 - \phi)$ haploid males

The first quantity, the proportion of diploids that are male, is the most useful one because if it can be estimated, then the frequency of matched matings and thus the number of sex alleles can be derived directly. However, the second quantity, the proportion of males that are diploid, is the more natural one to estimate. Operationally, it is the one most likely to be used, simply because when males are surveyed electrophoretically, for instance, a proportion of them turn out to be diploid. Unfortunately, it is the less informative of the two. Because one half of the progeny from a matched mating are diploid males, the frequency of diploids that are male is

$$\Phi = \theta(1 - s)/2$$

or

$$\Phi = (1 - s)/K$$

where s is the selection coefficient against the diploid males.

This represents the viability of diploid males from egg to adult and is different from the reduction in fitness incurred by the parents. It should be taken into account because diploid males in some species have low viability.[44]

Table 2 gives the proportion of diploids in a population expected to be male based on the second equation above. As can be seen from the top row ($s = 0$), even in a relatively large panmictic population, the proportion of diploids that are male will be around 4 to 5%, a biologically and statistically significant value. This represents the genetic load to the population as a result of the sex-determination locus. If diploid males have reduced viability, then, although they still will be produced as zygotes giving the same genetic load, the proportion in the adult population will be reduced, making their detection more difficult. In fact, reduced viability of diploid males will lead to an alteration in the observed sex ratio,[44] which will be discussed in Section V.

The quantity ϕ, the proportion of males that are diploid, depends not only on K, but also on the primary sex ratio, which I will define as the ratio of fertilized to unfertilized eggs. Table 3 gives values of ϕ as a function of K and the primary sex ratio for the case $s = 0$. If the sex ratio is even (1:1), then ϕ is close to, but less than Φ. However, if the primary sex ratio is biased, then ϕ varies considerably — from very low to quite high values. The point is that estimates of ϕ must be interpreted carefully, and some information on sex ratio is desirable in order to judge whether or not an estimate of ϕ is exceptionally "high", "low", or "normal". However, as I pointed out earlier, ϕ is a natural quantity to estimate, and now I will consider the problem of its estimation.

TABLE 2
The Proportion (Φ) of Diploids in a Population That are Male as a Function of K (the Number of Sex Alleles) and s (the Selection Coefficient Against the Diploid Males)

	K					
s	5	10	15	20	25	30
0	0.200	0.100	0.067	0.050	0.040	0.033
0.1	0.180	0.090	0.060	0.045	0.036	0.030
0.25	0.150	0.075	0.050	0.037	0.030	0.025
0.50	0.100	0.050	0.033	0.025	0.020	0.016
0.75	0.050	0.025	0.017	0.012	0.010	0.008
0.95	0.010	0.005	0.003	0.002	0.002	0.002

TABLE 3
The Proportion of Males in a Population That is Diploid (ϕ) as a Function of K (the Number of Sex Alleles) and the Primary Sex Ratio (the Ratio of Fertilized to Unfertilized Eggs) for the Case s = 0

	Sex ratio (fertilized:unfertilized eggs)		
K	3:1	1:1	1:3
5	0.375	0.167	0.062
10	0.231	0.091	0.032
15	0.167	0.063	0.022
20	0.130	0.048	0.016
25	0.107	0.038	0.013
30	0.090	0.032	0.011

C. MAXIMUM LIKELIHOOD ESTIMATES

Diploid males can be detected cytologically and morphologically; however, the most practical way to find diploid males, especially in species where there is no prior knowledge, is electrophoretically. Although there is a paucity of allozyme variation in the Hymenoptera, diploid males have "turned-up" during electrophoretic analysis of species carried out for other purposes.[40] Whether or not diploid males arise as a result of a single- or a multiple-locus Whiting scheme, they will of course be diploid at all loci (like females); thus, any heterozygote males can be identified unambiguously as diploid. Allozyme marker loci, which generally show codominance, are ideal for detecting diploid males. This method of estimation depends on taking a sample of males from a random-mating population (natural or laboratory).

Consider an allozyme locus with two alleles F and S at frequencies p and q, respectively, in both males and females. With random mating, the frequencies in the diploid and haploid offspring will be as shown in Figure 1. Considering only males then we have

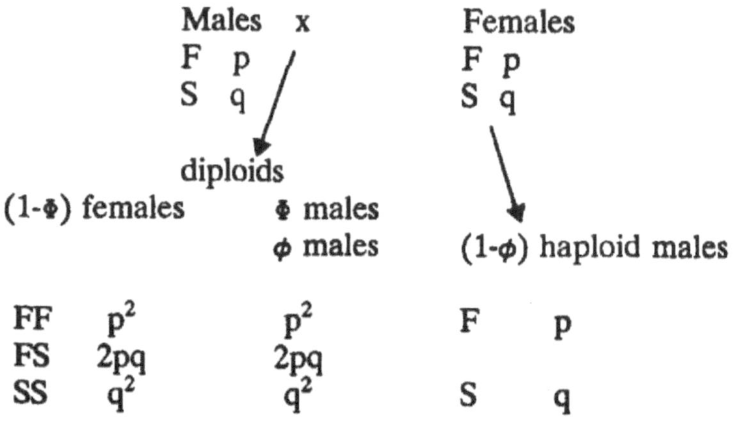

FIGURE 1. The expected frequency of genotypes at a marker locus with alleles *F* and *S* at frequencies p and q, respectively in both males and females. Random mating is assumed. Diploid males can be viewed in two ways: either as the proportion of diploids that are male (Φ) or as the proportion of males that are diploid (φ).

Number	A_2	+	B_2	+	C_2	=	T_2
Phenotype	*F*		*FS*		*S*		
Genotypes	*FF* + *F*		*FS*		*SS* + *S*		
Frequency	$\phi p^2 + (1 - f)p$		$2pq\phi$		$\phi q^2 + (1 - \phi)q$		

Assuming that only one offspring per random mating is included in the sample, then allele frequencies are given by

$$p = (A_2 + \tfrac{1}{2}B_2)/T_2, \quad q = (\tfrac{1}{2}B_2 + C_2)T_2$$

It is important to note that these are simple gene-counting estimates, which are the only estimates possible regardless of the proportion males that are diploid. Hence, these are maximum likelihood estimates. It follows that the maximum likelihood estimate of φ is given by

$$\phi = B_2 / 2pqT_2$$

with variance

$$V(\phi) = \phi[1 - 2pq\phi - (1 - 4pq)\phi^2] / 2pqT_2$$

These formulas and their derivation were given by Packer and Owen.[40,45] From the above considerations, it is useful to work out the sample size needed to detect a *single* heterozygous diploid male. This is given in Table 4 as a function of the parametric value of φ, and the allele frequency p at the marker locus. As can be seen with values of φ expected in reasonably large populations (i.e., around 5%) and with a marker locus with equal allele frequencies, *at least* 40 males must be scored to find a single heterozygote. This may not seem to be very many, but, in fact, routine population surveys of allozyme variation usually include numbers of males smaller than this.

TABLE 4
The Sample Size Required to Give an Expected Number of One Heterozygous Diploid Male, as a Function of ϕ (the Proportion of Males That Is Diploid) and p (the Allele Frequency at the Marker Locus)

	p					
ϕ	**0.05**	**0.1**	**0.2**	**0.3**	**0.4**	**0.5**
0.01	1053	556	313	239	209	200
0.02	527	278	157	120	105	100
0.05	211	112	63	48	42	40
0.10	106	56	32	24	21	20
0.20	53	28	16	12	11	10
0.30	36	19	11	8	7	7

When the allele frequencies at the marker locus are unequal or when values of ϕ are smaller, the sample size required increases considerably. Therefore, it is hardly surprising that diploid males have been detected in only a few species of parasitoids, although, in fact, we expect them to be relatively common.

With small values of ϕ, the variance will be large. This is illustrated by some data from Packer and Owen[40] on the halictine bee *Augochlorella striata*. The following numbers of male phenotypes were recorded at the Idh locus.

F	*FS*	*S*
80	1	25

giving $\phi = 0.0258 \pm 0.0256$ (S.E.). Clearly, when ϕ is small, the standard error is the same order of magnitude as the estimate.

However, the finding of a single diploid male does give valuable information and a valid estimate. The simple estimate (above) can only be calculated arithmetically for the case of a single marker locus with two alleles and using male data only. Owen and Packer[45] have extended this to include the cases of three alleles and two loci using male data only and two alleles with data from both males and females. Maximum likelihood estimates and their variances in these cases have to be obtained iteratively.

An alternative approach for multiple alleles and multiple loci, used by Kukuk and May,[39] is to combine data over all loci and estimate ϕ using an equation equivalent to

$$\phi = \left(\Sigma B_{ij}\right) / \left[1 - \Pi\left(\Sigma p_i^{\,2}\right)\right] T$$

where ΣB_{ij} is the sum of all heterozygous diploid males. This equation reduces to the single biallelic-locus equation given earlier, but, *except* in the single-locus case, it is not a maximum likelihood estimate, although it may be quite a good estimate.[45] The problem is that no variance can be attached to it, and the use of the binomial variance[39] *considerably underestimates* the true variance.

The estimate of ϕ, although interesting and valuable, does not on its own have a direct correspondence with K, the number of sex alleles, which is the quantity of most importance. On the other hand, Φ, the proportion of diploids that are male, does. A simple estimate of Φ can be obtained. Assume that a total of T *diploids* have been scored at a marker locus, and B_1 heterozygous females and B_2 heterozygous diploid males have been found. Then, from Figure 1, it can be seen that the ratio

$$B_2 / B_1 = \Phi 2pqT / (1 - \Phi) 2pqT$$

therefore, $\Phi = B_2/(B_1+B_2)$, with variance $V(\Phi) = \Phi(1 - \Phi)/(B_1+B_2)$, as given by Owen and Packer.[45]

There are two drawbacks with this method. One is that it assumes that a completely unbiased sample of male and female diploids is obtained, and if either are under-represented, then the estimate will be invalid. This may not be a problem in laboratory populations of parasitoids where, in principle at least, *all* individuals could be scored. A second problem is if diploid males have reduced viability, then this would lead to their underestimation. However, if diploid male mortality can be estimated independently (e.g., Petters and Mettus),[44] then a corrected estimate of Φ could be made. Assuming the above assumptions are met, then if

$$K = 1 / \Phi$$

Diploid males are expected to occur in most species of Hymenoptera, including parasitoids. Systematic surveys of populations using allozyme marker loci will detect diploid males if a sufficient sample size is used. A simple formula gives a maximum likelihood estimate of ϕ, but some caution must be used in its biological interpretation.

D. DIPLOID MALE MORPHOLOGY AND VIABILITY

Apart from their genotype at the sex locus or loci, diploid males are, of course, genetically the same as their sisters. Therefore, because sexual size dimorphism is common in the Hymenoptera, it is not surprising that diploid males differ morphologically from haploid males. Diploid males of the fire ant *Solenopsis invicta* are significantly larger on average than haploid males.[36] Similarly, the single diploid male found in *A. striata* was significantly larger than average.[40] Naito and Suzuki[34] were also able to discriminate haploid and diploid males of the parasitoid *Athalia rosae* on the basis of size. Diploid males of *Bracon hebetor* are larger than haploid males for many morphological characters,[46] and Smith and Wallace[33] found that pupal weights of diploid males of *Neodiprion nigroscutum* were 20 to 30% heavier than their haploid brothers.

Importantly, diploid males not only differ from haploid males morphologically, but also from females. Grosch[46] found significantly different frequency distributions of microchaetae per square millimeter in *B. hebetor*. The mean number was lower in diploid males than females, suggesting that cell size in diploid males is larger than in females. Similarly, in *N. nigroscutum*, the pupal weight distributions of diploid males and females are significantly different.[33] These observations suggest that development is altered in diploid males, and this would account for the low viability observed in some species. Petters and Mettus[44] found significantly lower hatchability of diploid males on caterpillars in *B. hebetor*. On the other hand, a study of the adult lifespan of *Habrobracon serinopae* males showed no difference between diploid and haploid male longevity, in spite of a larger cell size in the diploid males.[47]

V. SEX RATIO AND MATING SYSTEMS

The sex ratio in populations of parasitoids used for biological control is obviously important. All else being equal, the optimal sex ratio in a panmictic population of haplodiploids is 1:1.[48] However, the observed sex ratio in natural or laboratory populations of parasitoid Hymenoptera is often female or male biased and is the result of either environmental or genetic causes. Thus, potentially, sex ratios of species could be manipulated (e.g., toward females) to provide more effective control. Environmental variables such as temperature, humidity, photoperiod, and host size have been shown to influence offspring sex ratios in many parasitoid wasps.[25] Here, I will concentrate on genetic factors.

A. LOCAL MATE COMPETITION AND INBREEDING

Hamilton[49] pointed out that in structured populations with nonrandom mating, the optimal offspring sex ratio is not 1:1, but is female biased. The type of population structure that favors this is found in many parasitoids. For instance, in the gregarious parasitoid *Nasonia vitripennis*, one or more foundress females will lay eggs on a single host puparium. After eclosion, the offspring mate amongst themselves as a group before the females disperse to find new groups, and thus, there is *local mate competition* or LMC.[50] In this species, the males are wingless. If a single female lays eggs in a patch, then she need only produce enough males to fertilize her daughters, giving a highly female-biased offspring ratio. As the number of foundresses increases, mating becomes more and more population wide, and the optimal ratio approaches 1:1. This can be analyzed as a genetic polymorphism;[51] however, in reality, it is manifest as behavioral flexibility on the part of female wasps. Thus, any female is capable of laying any sex ratio of eggs, but adjusts the ratio facultatively according to the perceived conditions. This is made possible by the haplodiploid genetic system, as a female is able to determine whether an egg she lays is fertilized or not by controlling the release of sperm from her spermatheca. The optimal offspring sex ratio is approximately given by $r = (N - 1)/(1 + G)N$, where r is the proportion of males, N is the number of foundress females, and G is the relative genetic relatedness of a foundress to her daughters compared to her sons.[50]

G depends on the degree of inbreeding, hence $G = 2/(2 - S)$, where S is the probability of sib-mating.[50,52] Experiments on *N. vitripennis*[53] and fig wasps[54,55] have supported the theoretical relationship between foundress number and r. However, in fig wasps, where species with S ranging from 0.34 to 0.91 have been studied, there is not such a clear-cut relationship between inbreeding and sex ratio.[50] Similarly, the degree of inbreeding of a foundress in *Leptopilina heterotoma*, a *Drosophila* parasitoid, does not have an effect on the sex ratio of her progeny.[56] Although wasps show facultative sex-ratio adjustment, some genetic variation for sex ratio does occur also[57] apart from that caused by extrachromosomal factors (see Section VII).

B. INBREEDING, SEX RATIOS, AND DIPLOID MALES

Species like *N. vitripennis* (with wingless males) and fig wasps tend to be obligate inbreeders, and therefore, it is not surprising that their sex-determination system is such that diploid males are not easily produced.[37] However, other species which are normally outbred can also adjust their offspring sex ratios facultatively.[58] Similarly, environmental factors may be responsible for biased sex ratios in many of these species.[25] Clearly, in these types of species, the occurrence of diploid males could have a significant effect on the observed sex ratio. This will be particularly severe if species are forced, either naturally or artificially, to inbreed. For instance, Biemont and Bouletreau[59] inbred French and Guadeloupe *Cothonaspis boulardi*, a parasitoid of *Drosophila*, and found a significant increase in the percentage of males in the F_1 offspring. They concluded that this was probably due to diploid male production, since there was no decrease in productivity of the lines and no inbreeding effects caused by the increase in overall homozygosity. The low levels of genetic variation in Hymenoptera may ameliorate the effects of inbreeding per se.

The effect that diploid males have on the observed adult sex ratio in a population is given in Table 5. It is assumed that diploid and haploid males are, or cannot be, distinguished and are combined to give the observed ratio. This depends greatly on the viability of the diploid males. Even in populations with a high number of sex alleles, ratios can be significantly biased if diploid male viability is high. In inbred laboratory populations, the effect will be the most severe. Male-biased ratios are common in laboratory cultures of parasitoids, and diploid males can be added to the list of possible causes given by Waage.[60] One other factor, discussed in the Section V.C, is the higher mortality rate expected in *haploid* males as a result of lethal mutations.[61] This has the effect of shifting the observed ratio back toward the primary ratio, although in many cases the sex ratio remains significantly biased (Table 5).

C. MUTATION AND HAPLOID MALE MORTALITY

Haploid eggs are expected to have a higher mortality rate than diploid eggs because there is no masking of recessive lethal alleles arising through mutation.[61] Smith and Shaw[61] suggest that this will have an underlying affect on all haplodiploid sex ratios. The probability of a haploid male dying because of a new mutation or from a recessive mutation inherited from his mother is $P = 1 - \exp(-3m)$, where m is the mutation rate per genome per generation.[61] Using data on survival of broods of *Apanteles* spp., Smith and Shaw[61] estimate m to be about 0.035, giving a differential haploid mortality of 10%.

There is no doubt that the haploid state is very vulnerable to mutation. Irradiated adult haploid males of *H. serinopae* have a significantly shorter lifespan than do diploid males.[47]

Haploid male mortality is also another factor influencing sex-ratio adjustment in species always experiencing highly female-biased sex ratios through LMC. If a single foundress female is present in a patch, then she needs only to produce a single

TABLE 5
The Observed Sex Ratio (F:M) as a Function of K, s, Haploid Male Mortality, and the Primary Sex Ratio (Fertilized:Unfertilized Eggs)

		Haploid male mortality:			
		0%		10%	
		Primary sex ratio			
K	s	3:1	1:1	3:1	1:1
5	0	1.50:1	1:1.50	1.60:1	1:1.37
	0.25	1.65:1	1:1.44	1.78:1	1:1.31
	0.50	1.85:1	1:1.37	2.00:1	1:1.25
	0.95	2.33:1	1:1.26	2.58:1	1:1.14
10	0	2.08:1	1:1.22	2.25:1	1:1.11
	0.25	2.20:1	1:1.19	2.40:1	1:1.08
	0.50	2.35:1	1:1.17	2.57:1	1:1.06
	0.95	2.66:1	1:1.12	2.95:1	1:1.01
20	0	2.48:1	1:1.10	2.71:1	1:1.00
	0.25	2.56:1	1:1.09	2.81:1	1:0.95
	0.50	2.65:1	1:1.08	2.92:1	1:0.97
	0.95	2.83:1	1:1.05	3.14:1	1:0.95

Note: In the observed sex ratio, it is assumed that diploid and haploid males are not distinguished and are counted together.

male to fertilize her daughters.[49] However, if she lays just a single unfertilized egg, there is a probability that it will not survive.[62] Moreover, if, while laying a clutch of eggs, she fertilizes the eggs at random, there is a chance that in small broods no unfertilized eggs will be laid.[60,62] Therefore, to ensure the presence of at least one adult male in each brood, a higher sex ratio than predicted by LMC alone and a more precise, nonrandom pattern of sex allocation is expected.[62,63] In species such as *Goniozus gordhi* and *G. emigratus* that normally lay small, female-biased clutches, precise sex ratios are indeed found. On the other hand, in normally outbred species such as *Asobara persimilis*, the number of males in a brood is determined binomially.[58] Some methods of estimating primary sex ratio and progeny survival are given by Wellings, Morton, and Hart.[64]

D. MATING SYSTEMS

In species susceptible to diploid male production, selection should act to promote outcrossing. *B. hebetor* females lay gregarious broods that are female biased,[2] but sib-mating of offspring would be highly deleterious. Antolin and Strand[65] studied a field population of *B. hebetor* and found, however, that the postemergence behavior of males and females lead to outcrossing. Most females dispersed from their emergence site without mating, and males aggregated on high places and waited for females.

Multiple mating reduces the expected proportion of diploid males per brood,[66] but the proportion of broods in the population producing diploid males increases.[67] Multiple mating by females is quite common in the Hymenoptera and has been confirmed by observation or using genetic markers in at least four parasitoids.[65,68]

VI. GEOGRAPHICAL VARIATION, HOST-PARASITE COEVOLUTION, AND SEXUAL DIMORPHISM

Geographical variation in response to abiotic factors, host suitability, and in resistance to toxins has been found in parasitoids.[30] The exact genetic basis is unknown, but it is probably polygenic in nature.

Parasitoids in the genus *Leptopilina* of different geographic origins vary in their parasitization success on different host populations of *Drosophila*.[69] In *L. heterotoma*, a single codominant gene inhibits encapsulation by the host.[69]

It is important to distinguish between direct interactions between parasitoids and their hosts and the reactions of parasitoids to the habitats of their hosts.[69] Evolutionary changes in both host and parasitoid have been documented in experimental systems (e.g., Pimentel, Nagel, and Madden),[70] and it is clear that the searching behavior of parasitoids may be responsible for the maintenance of a larval locomotory polymorphism in *D. melanogaster*.[71]

Parasitoids show a response to environmental factors parallel to that of their hosts. A good example is tolerance and utilization of ethanol by *Drosophila* parasitoids.[72,73] *Drosophila* adults that lay eggs on fermenting fruit tolerate relatively high (15 to 20%) concentrations of atmospheric ethanol,[72] and they utilize low concentrations as a metabolic resource.[74] A parallel response is found in most larval parasitoids of *Drosophila*, but not in pupal parasitoids; however, the degree of tolerance is less than that in *D. melanogaster*.[72] An interesting aspect is that female parasitoids are significantly more tolerant to ethanol than males.[72,73] This is not the case with *Drosophila*.

This is not a sex-limited expression, since the males do show significant tolerance and utilization.[73] In *Drosophila*, alcohol tolerance depends to some extent on alcohol dehydrogenase (ADH) activity. If a similar detoxification system operates in the parasitoids, then higher ADH activity in females (if it occurs) could be explained by haplodiploidy through lack of dosage compensation at this locus. However, there is no evidence for this.[72] On the other hand, sex differences in enzyme activity are known in parasitoids. Male *N. vitripennis* have significantly lower specific activity of glycerolphosphate dehydrogenase than females.[75] This is an important enzyme in the glycolytic metabolism of flight muscles; therefore, the difference in activity between the wingless males and the winged females is understandable.[75] The situation is not consistent in *B. hebetor*, where some enzymes show higher activity in females and others show higher activity in males.[2]

VII. EXTRACHROMOSOMAL INHERITANCE

Instances of extranuclear inheritance have been described in parasitoids. Sex-ratio distortion occurs in *N. vitripennis* by three extrachromosomal factors: (1) son killer, a maternally inherited bacterium passed on in the cytoplasm which kills male eggs; (2) paternal sex ratio, a "selfish" B chromosome which eliminates the paternal genome and is transmitted by sperm, causing condensation and loss of paternal chromosomes in fertilized eggs; and (3) maternal sex ratio, maternally inherited and causing 97% female families.[76,77]

Muscidifurax raptorellus exhibits polygenically controlled oviposition behavior. Oviposition is either solitary or gregarious, with more than one individual developed per host. In addition to the polygenic system, there are extranuclear influences which cause changes in the oviposition phenotype of females.[22] Males change a female's oviposition phenotype upon mating, and this appears to be due to the transmission of an (as yet) unknown substance.[21] Females with the solitary genotype express gregarious oviposition after mating with a male of gregarious genotype. Conversely, the gregarious behavior of females is reduced after mating with males of the solitary genotype.[22]

VIII. CONCLUSIONS

Parasitoids have levels of allozyme variation which are low, yet comparable to those found in other Hymenoptera, and which probably result from haplodiploidy. There is strong selection at the sex determination locus or loci in many, usually outbred species. However, in spite of this, diploid males are expected to occur in most populations of parasitoids at frequencies of around 5% and represent an inevitable genetic load. They are likely to be a problem in laboratory populations founded with a small number of individuals. Diploid males can be detected and their proportion estimated using marker loci such as allozymes. In fact, they have only been found in a few species, and this is probably because insufficient sample sizes have been used in routine surveys of allozyme variation. Diploid males can have a significant biasing effect on the observed sex ratio and should be accounted for if possible.

Sex ratios of parasitoids are influenced by many environmental and genetic factors; however, a major characteristic is that females of many species are able to adjust their offspring sex ratios facultatively. At a proximate level, this is due to haplodiploidy and the fact that females control whether or not they fertilize an egg. Both haploid and diploid male mortality influence the primary and adult sex ratios in this case.

ACKNOWLEDGMENTS

I am most grateful to Dr. S. K. Narang for inviting me to contribute to the workshop and to Dr. William Steiner for his comments on the manuscript. I thank the Natural Sciences and Engineering Research Council of Canada, Ottawa for research funding and Mount Royal College Calgary, Alberta, Canada for travel funds.

REFERENCES

1. **Greathead, D. J.,** Parasitoids in classical biological control, in *Insect Parasitoids*, Waage, J. and Greathead, D., Eds., Academic Press, London, 1986, 290.
2. **Grosch, D. S.,** Genetics research on braconid wasps, *Adv. Genet.*, 25, 109, 1988.
3. **Cassidy, J. D.,** The parasitoid wasps *Habrobracon* and *Mormoniella*, in *Handbook of Genetics*, Vol. 3, King, R. C., Ed., Plenum Press, New York, 1975, 173.
4. **Saul, G. B., Saul, S. W., and Becker, S.,** Linkage in *Mormoniella*, *Genetics*, 57, 369, 1967.
5. **Graur, D.,** Gene diversity in Hymenoptera, *Evolution*, 39, 190, 1985.
6. **Owen, R. E.,** Difficulties with the interpretation of patterns of genetic variation in the eusocial Hymenoptera, *Evolution*, 39, 201–205, 1985.
7. **Crespi, B. J.,** Heterozygosity in the haplodiploid Thysanoptera, *Evolution*, 45, 458, 1991.
8. **Avery, P. J.,** The population genetics of haplo-diploids and X-linked genes, *Genet. Res.*, 44, 321, 1984.
9. **Pamilo, P.,** Genic variation at sex-linked loci: quantification of regular selection models, *Hereditas*, 91, 129, 1979.
10. **Curtsinger, J. W.,** On the opportunity for polymorphism with sex-linkage or haplodiploidy, *Genetics*, 96, 995, 1980.
11. **Pamilo, P. and Crozier, R. H.,** Genic variation in male haploids under deterministic selection, *Genetics*, 98, 199, 1981.
12. **Owen, R. E.,** Selection at two sex-linked loci, *Heredity*, 60, 415, 1988.
13. **Crozier, R. H.,** Counter-intuitive property of effective population size, *Nature (London)*, 262, 384, 1976.
14. **Nagylaki, T.,** The inbreeding effective population number and the expected heterozygosity for an X-linked locus, *Genetics*, 47, 731, 1981.
15. **Crow, J. F. and Kimura, M.,** *An Introduction to Population Genetics Theory*, Harper and Row, New York, 1970.
16. **Unruh, T. R., White, W., Gonzalez, D., Gordh, G., and Luck, R. F.,** Heterozygosity and effective size in laboratory populations of *Aphidius ervi* [*Hym.: Aphidiidae*], *Entomophaga*, 28, 245, 1983.
17. **Hassel, M. P.,** Parasitoids and population regulation, in *Insect Parasitoids*, Waage, J. and Greathead, D., Eds., Academic Press, London, 1986, 201.
18. **Wright, S.,** *Evolution and Genetics of Populations. Vol. 2. The Theory of Gene Frequencies*, University of Chicago Press, Chicago, 1969.
19. **Eickwort, K.R.,** Differential variation of males and females in *Polistes exclamans*, *Evolution*, 23, 391, 1969.
20. **Owen, R.E.,** Differential size variation of male and female bumblebees, *J. Hered.*, 80, 39, 1989.
21. **Legner, E.F.,** Inheritance of gregarious and solitary oviposition in *Muscidifurax raptorellus* Kogan and Legner (Hymenoptera: Pteromalidae), *Can. Entomol.*, 119, 791, 1987.
22. **Legner, E. F.,** Estimations of number of active loci, dominance and heritability in polygenic inheritance of gregarious behavior in *Muscidifurax raptorellus* [*Hymenoptera: Pteromalidae*], *Entomophaga*, 36, 1, 1991.
23. **Legner, E. F.,** *Muscidifurax raptorellus* (Hymenoptera: Pteromalidae) females exhibit postmating behavior typical of the male genome, *Ann. Entomol. Soc. Am.*, 81, 522, 1988.
24. **Sequeira, R. and Mackauer, M.,** Quantitative genetics of body size and development time in the parasitoid wasp *Aphidius ervi* (Hymenoptera: Aphidiidae), *Can. J. Zool.*, 70, 1102, 1992.
25. **King, B. H.,** Offspring sex ratios in parasitoid wasps, *Q. Rev. Biol.*, 62, 367, 1987.
26. **Wilkes, A.,** The effects of selective breeding on the laboratory propagation of insect parasites, *Proc. R. Soc. London Ser. B*, 134, 227, 1947.
27. **Allen, H. W.,** Propagation of *Horogenes molestae*, an Asiatic parasite of the Oriental fruit moth on the potato tuberworm, *J. Econ. Entomol.*, 47, 278, 1954.
28. **White, E. B., DeBach, P., and Garber, M. J.,** Artificial selection for genetic adaptation to temperature extremes in *Aphytis linganensis* Compere (Hymenoptera: Aphelinidae), *Hilgardia*, 40, 161, 1970.

29. **van Lentern, J. C.,** Parasitoids in the greenhouse: successes with seasonal inoculative release systems, in *Insect Parasitoids*, Waage, J. and Greathead, D., Eds., Academic Press, London, 1986, 342.
30. **Unruh, T. R. and Messing, R. H.,** Intraspecific biodiversity in Hymenoptera: implications for conservation and biological control, in *Hymenoptera and Biodiversity*, LaSalle, J. and Gauld, I., Eds., C.A.B. International, London, 1993, 368.
31. **Whiting, P. W.,** Multiple alleles in complementary sex determination of *Habrobracon*, *Genetics*, 28, 365, 1943.
32. **Camargo, C. A.,** Sex determination in bees. XI. Production of diploid males and sex determination in *Melipona quadrifasciata*, *J. Apic. Res.*, 18, 77, 1979.
33. **Smith, S. G. and Wallace, D. R.,** Allelic sex determination in a lower hymenopteran, *Neodiprion nigroscutum* Midd., *Can. J. Genet. Cytol.*, 13, 617, 1971.
34. **Naito, T. and Suzuki, H.,** Sex determination in the sawfly, *Athalia rosae ruficornis* (Hymenoptera): occurrence of triploid males, *J. Hered.*, 82, 101, 1991.
35. **Hedderwick, M. P., El Agoze, M., and Periquet, G.,** Enzymatic polymorphism, inbreeding and occurence of diploid males in *Diadromus pulchellus*, Les Colloques d l'INRA, 48, 109, 1988.
36. **Crozier, R. H.,** Heterozygosity and sex determination in haplo-diploidy, *Am. Nat.*, 105, 399, 1971.
37. **Skinner, S. W. and Werren, J. H.,** The genetics of sex determination in *Nasonia vitripennis* (Hymenoptera, Pteromalidae), *Genetics*, 94, (Suppl.), 98, 1980.
38. **Ross, K. G. and Fletcher, D. J. C.,** Genetic origin of male diploidy in the fire ant *Solenopsis invicta* (Hymenoptera: Formicidae), and its evolutionary significance, *Evolution*, 39, 888, 1985.
39. **Kukuk, P. F. and May, B.,** Diploid males in a primitively eusocial bee, *Lasioglossum (Dialictus) zephyrum* (Hymenoptera: Halictidae), *Evolution*, 44, 1522, 1990.
40. **Packer, L. and Owen, R. E.,** Allozyme variation, linkage disequilibrium and diploid male production in a primitively social bee *Augochlorella striata* (Hymenoptera: Halictidae), *Heredity*, 65, 241, 1990.
41. **Adams, J., Rothman, E. D., Kerr, W. E., and Paulino, Z. L.,** Estimation of the number of sex alleles and queen matings from diploid male frequencies in a population of *Apis mellifera*, *Genetics*, 86, 583, 1977.
42. **Woyke, J.,** Population genetic studies on sex alleles in the honeybee using the example of the Kangaroo Island bee sanctuary, *J. Apic. Res.*, 15, 105, 1976.
43. **Yokoyama, S. and Nei, M.,** Population dynamics of sex-determining alleles in honey bees and self-incompatibility alleles in plants, *Genetics*, 91, 609, 1979.
44. **Petters, R. M. and Mettus, R. V.,** Decreased diploid male viability in the parasitic wasp, *Bracon hebetor*, *J. Hered.*, 71, 353, 1980.
45. **Owen, R. E. and Packer, L.,** Estimation of the proportion of diploid males in populations of Hymenoptera, 1993, 73.
46. **Grosch, D. S.,** The relation of cell size and organ size to mortality in *Habrobracon*, *Growth*, 9, 1, 1945.
47. **Clark, A. M., Bertrand, H. A., and Smith, R. E.,** Life span differences between haploid and diploid males of *Habrobracon serinopae* after exposure as adults to X rays, *Am. Nat.*, 97, 203, 1963.
48. **Hartl, D. L. and Brown, S. W.,** The origin of male haploid genetic systems and their expected sex ratio, *Theor. Popul. Biol.*, 1, 165, 1970.
49. **Hamilton, W. D.,** Extraordinary sex ratios, *Science*, 156, 477, 1967.
50. **Werren, J. H.,** Labile sex ratios in wasps and bees, *Bioscience*, 37, 498, 1987.
51. **Taylor, P. D. and Bulmer, M. G.,** Local mate competition and the sex ratio, *J. Theor. Biol.*, 86, 409, 1980.
52. **Suzuki, Y. and Iwasa, Y.,** A sex ratio theory of gregarious parasitoids, *Res. Popul. Ecol.*, 22, 366, 1980.
53. **Werren, J. H.,** Sex ratio evolution under local mate competition in a parasitic wasp, *Evolution*, 37, 116, 1983.
54. **Frank, S. A.,** Hierarchical selection theory and sex ratios. II. On applying the theory, and a test with fig wasps, *Evolution*, 39, 949, 1985.

55. **Herre, E. A.,** Sex ratio adjustment in fig wasps, *Science*, 22, 896, 1985.
56. **Hey, J. and Gargiulo, M. K.,** Sex-ratio changes in *Leptopilina heterotoma* in response to inbreeding, *J. Hered.*, 76, 209, 1985.
57. **Parker, E. D. and Orzack, S. H.,** Genetic variation for the sex ratio in *Nasonia vitripennis*, *Genetics*, 110, 93, 1985.
58. **Owen, R. E.,** Sex ratio adjustment in *Asobara persimilis* (Hymenoptera: Braconidae), a parasitoid of *Drosophila*, *Oecologia*, 59, 402, 1983.
59. **Biemont, C. and Bouletreau, M.,** Hybridization and inbreeding effects on genome coadaption in a haplo-diploid hymenoptera: *Cothonaspis boulardi* (Eucoilidae), *Experientia*, 36, 45, 1980.
60. **Waage, J. K.,** Family planning in parasitoids: adaptive patterns of progeny and sex allocation, in *Insect Parasitoids*, Waage, J. and Greathead, D., Eds., Academic Press, London, 1986, 63.
61. **Smith, R. H. and Shaw, M. R.,** Haplodiploid sex ratios and the mutation rate, *Nature (London)*, 287, 728, 1980.
62. **Hartl, D. L.,** Some aspects of natural selection in arrhenotokous populations, *Am. Zool.*, 11, 309, 1971.
63. **Green, R. F., Gordh, G., and Hawkins, B. A.,** Precise sex ratios in highly inbred parasitic wasps, *Am. Nat.*, 120, 653, 1982.
64. **Wellings, P. W., Morton, R., and Hart, P. J.,** Primary sex-ratio and differential progeny survivorship in solitary haplo-diploid parasitoids, *Ecol. Entomol.*, 11, 341, 1986.
65. **Antolin, M. F. and Strand, M. R.,** Mating systems of *Bracon hebetor* (Hymenoptera: Braconidae), *Ecol. Entomol.*, 17, 1, 1992.
66. **Page, R. E.,** The evolution of multiple mating behavior by honey bee queens (*Apis mellifera* L.), *Genetics*, 96, 263, 1980.
67. **Pamilo, P.,** Evolution of colony characteristics in social insects. II. Number of reproductive individuals, *Am. Nat.*, 138, 412, 1991.
68. **Page, R. E. and Metcalf, R. A.,** Multiple mating, sperm utilization, and social evolution, *Am. Nat.*, 119, 263, 1982.
69. **Bouletreau, M.,** The genetic and coevolutionary interactions between parasitoids and their hosts, in *Insect Parasitoids*, Waage, J. and Greathead, D., Eds, Academic Press, London, 1986, 169.
70. **Pimentel, D., Nagel, W. P., and Madden, J. L.,** Space-time structure of the environment and the survival of parasite-host systems, *Am. Nat.*, 97, 141, 1963.
71. **Sokolowski, M. B. and Turlings, T. C. J.,** *Drosophila* parasitoid-host interactions: vibrotaxis and ovipositor searching from the host's perspective, *Can. J. Zool.*, 65, 461, 1987.
72. **Bouletreau, M. and David, J. R.,** Sexually dimorphic response to host habitat toxicity in *Drosophila* parasitic wasps, *Evolution*, 35, 395, 1981.
73. **Owen, R. E.,** Utilization and tolerance of ethanol, acetic acid and acetaldehyde vapour by *Asobara persimilis*, a parasitoid of *Drosophila*, *Entomol. Exp. Appl.*, 39, 143, 1985.
74. **Parsons, P. A., Stanley, S. M., and Spence, G. E.,** Environmental ethanol at low concentrations: longevity and development in the sibling species *Drosophila melanogaster* and *D. simulans*, *Aust. J. Zool.*, 27, 747, 1979.
75. **McGuffin, M. E. and Collier, G. E.,** Developmental and biochemical differentiation of glycerolphosphate dehydrogenase isozymes in *Mormoniella vitripennis* (Hymenoptera: Pteromalidae), *Insect Biochem.*, 12, 405, 1982.
76. **Werren, J. H., Skinner, S. W., and Huger, A. M.,** Male-killing bacteria in a parasitic wasp, *Science*, 231, 990, 1986.
77. **Nur, U., Werren, J. H., Eickbush, D. G., Burke, W. D. and Eickbush, T. H.,** A "selfish" B chromosome that enhances its transmission by eliminating the paternal genome, *Science*, 240, 512, 1988.
78. **Lester, L. J. and Selander, R. K.,** Population genetics of haplodiploid insects, *Genetics*, 92, 1329, 1979.
79. **Menken, S. B. J.,** Enzymatic characterization of nine endoparasitic species of small ermine moths (Yponomeutidae), *Experientia*, 38, 1461, 1982.

80. **Propp, G. D.,** Characterization of intra- and interspecific isoenzyme variation in *Spalangia* spp. and *Muscidifurax* spp. (Chalcidoidea: Pteromalidae), parasitoids of synanthropic Diptera, *Misc. Publ. Entomol. Soc. Am.*, 61, 164, 1986.
81. **Hung, A. C. F., Day, W. H., and Hedlund, R. C.,** Genetic variability in arrhenotokous and thelytokous forms of *Mesochorus nigripes* (Hymenoptera: Ichneumonidae), *Entomophaga*, 33, 7, 1988.
82. **Unruh, T. R., White, W., Gonzalez, D., and Woolley, J. B.,** Genetic relationships among seventeen *Aphidius* (Hymenoptera: Aphidiidae) populations, including six species, *Ann. Entomol. Soc. Am.*, 82, 754, 1989.
83. **Shaumar, N., Rojas-Rouse, D., and Pasteur, N.,** Allozyme polymorphism in the parasitic hymenoptera *Diadromus pulchellus* WSM (Ichneumonidae), *Genet. Res.*, 32, 47, 1978.
84. **Steiner, W. W. M. and Teig, D. A.,** *Microplitis croceipes* (Cresson): genetic characterization and developing insecticide resistant biotypes, *Southwest. Entomol.*, Suppl. 12, 71, 1989.

Chapter 5

LABORATORY SELECTION FOR PESTICIDE RESISTANCE IN NATURAL ENEMIES

Marshall W. Johnson and Bruce E. Tabashnik

TABLE OF CONTENTS

0-8493-2607-9/94/$0.00+$.50

I. INTRODUCTION

Many arthropod pests targeted in biological control programs inhabit agroecosystems with numerous other pest species for which biological control is not effective. Introduction and establishment of effective natural enemies may not solve a pest problem if provisions are not taken to protect biological control agents from pesticide applications directed at those pests which have no effective alternative controls. In these cases, secondary pest outbreaks may occur. When pesticides are applied for pests with effective natural enemies, pest resurgence may occur due to decimation of the natural enemies. When natural enemies are more susceptible to pesticides than their hosts, the probability of resurgence increases. Biological control under these conditions may be improved by modification of environmental factors adverse to biological control (conservation) or an increase in the number or improvement of the fitness of natural enemies (augmentation).[1]

Conservation may fail due to the lack of selective pesticides to which biological control agents are naturally tolerant[2] and the ineffectiveness of alternative controls for the pests without effective natural enemies. Augmentation includes (1) mass rearing and release of natural enemies on either an innoculative or inundative basis and (2) genetic modification of natural enemies to improve specific traits (e.g., pesticide resistance, temperature tolerance, and searching abilities). If natural enemies cannot be conserved due to deleterious pesticide applications, then mass releases of susceptible natural enemies will not usually provide effective biological control. Laboratory selection for pesticide resistance in natural enemies is one option that has been used to circumvent this problem.[3,4] Beckendorf and Hoy,[5] Croft,[6] and Hoy[7,8] have provided excellent reviews of the achievements of selection programs.

We present a modified version of Hoy's[9] flow chart (Figure 1) that outlines 15 general steps in a selection program. The steps can be divided into four major phases: (Phase I) determination of needs (steps a–b), (Phase II) selection for resistance (steps c–f), (Phase III) postselection evaluation (steps g–i), and (Phase IV) release and establishment of selected strains (steps k–o).

This chapter reviews the progress of genetic selection of natural enemies for resistance to pesticides. Specifically, we (1) review the success of selection programs with respect to steps outlined in the flow chart, (2) identify bottlenecks in the process that limit success, and (3) use estimates of realized heritability to evaluate the selection process.

II. SUCCESS OF SELECTION PROGRAMS

Several predatory arthropod species have undergone laboratory selection for pesticide resistance. Phytoseiid mite species in this group include *Phytoseiulus persimilis* Athias-Henriot,[10-12] *Amblyseius fallacis* (Garman),[13] *A. nicholsi* Ehara and Lee,[14] *Metaseiulus occidentalis* (Nesbitt),[15,16] and *Typhlodromus pyri* Scheuten.[11] *M. occidentalis* has been selected for resistance to carbaryl, dimethoate,[15] permethrin,[17]

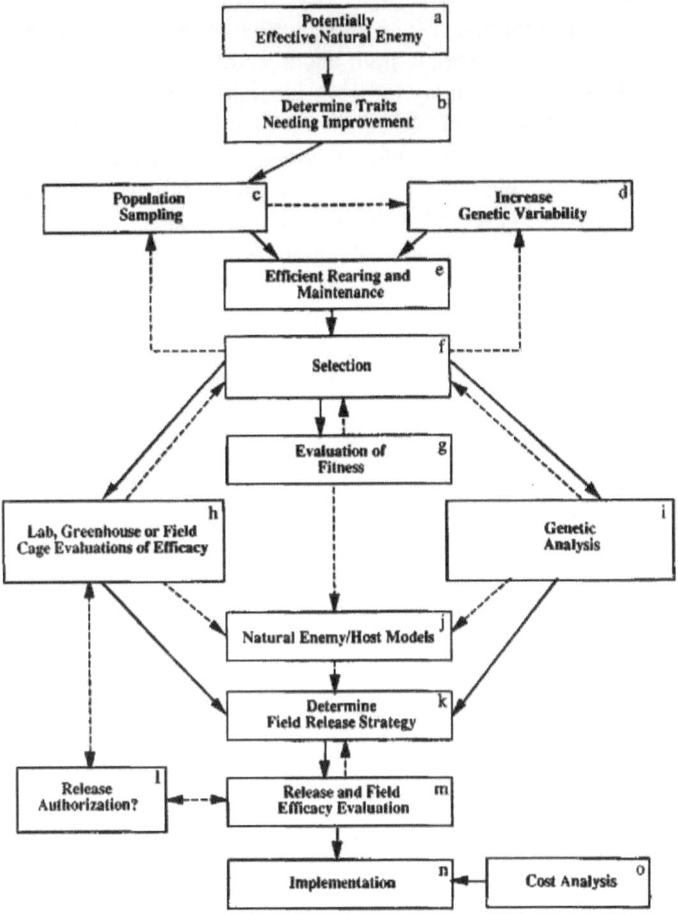

FIGURE 1. Program flow chart for genetic selection and implementation of a pesticide-resistant natural enemy strain. (Modified from Hoy, M. A., *Recent Advances in Knowledge of the Phytoseiidae*, Division of Agricultural Science, Publ. 3284, University of California, Berkeley, 1982, 72.)

and abamectin.[16] Selected strains of *M. occidentalis* are produced commercially for mass releases in apple orchards in the state of Washington and in British Columbia, Canada, as well as in almond orchards in California.[18,19] An insect predator, the common green lacewing *Chrysoperla carnea* (Stephens) (Neuroptera: Chrysopidae), has also been selected for carbaryl resistance.[20,21]

Studies conducted using parasitoids in selection programs[22-28] have had limited success compared with that for predators.[29] Although moderate responses to selection have been obtained, the selection response has not been adequate to incorporate the selected strain into an integrated pest management (IPM) program. Selection conducted on colonies established from field populations of *Aphytis melinus* DeBach, a parasitoid of the California red scale exhibiting low levels of resistance to various insecticides, produced a parasitoid strain about 12-fold more resistant to carbaryl than a standard susceptible colony.[30] Using laboratory selection, azinphosmethyl (Guthion®)* resistance was increased 7.5-fold in the parasitoid *Trioxys pallidus* Haliday, which attacks the walnut aphid, *Chromaphis juglandicola* Kaltenbach, in California.[31] Although azinphosmethyl resistance was significantly increased, bioassays suggested that the selected *T. pallidus* strain could not survive rates of azinphosmethyl applied in the field. Laboratory studies showed that the selected azinphosmethyl-resistant *T. pallidus* strain exhibited cross-resistance to chlorpyrifos (Lorsban®),** endosulfan (Thiodan®),*** methidathion (Supracide®),† and phosalone (Zolone®)‡.[31]

Experimental field studies showed that the azinphosmethyl-resistant *T. pallidus* strain could establish and persist in walnut orchards receiving one to two seasonal applications of azinphosmethyl and methidathion.[32] Establishment and persistence of selected *T. pallidus* in the field studies were greater than expected, based on estimates of survival from laboratory bioassays alone. These results demonstrated that critical evaluation of fitness for a selected strain should be made under the field conditions in which the selected strain must ultimately function.

The examples above illustrate the few successes in the development of pesticide-resistant natural enemies. One may ask why there have not been more successes. To provide insights into this problem, we reviewed 16 selection programs for predators and 22 programs for parasitoids to determine which steps (Figure 1) were completed in each program.

A. PHASE I: DETERMINATION OF NEEDS

The initial steps in all selection programs are identification of a potentially effective natural enemy (Figure 1, step a) and determination of the trait or traits needing improvement (Figure 1, step b). Each selection project completed these two steps (Table 1). These steps are necessary and to some degree determine the outcome of the program.

All predators except one (*Chrysoperla carnea*) used in selection studies were in the Acari: Phytoseiidae. All parasitoids were in the order Hymenoptera, with the most attention directed toward the genus *Aphytis* (Aphelinidae). Most natural enemies in these programs were oligophagous [*Aphytis melinus* DeBach, *A. holoxanthus* DeBach, *Encarsia formosa* (Gahan), *Bracon mellitor* (Say), *T. pallidus* Haliday, *Amblyseius fallacis*, and *M. occidentalis*]; only one was a generalist (*C. carnea*). Selection for resistance involved several compounds including chlorinated hydrocarbons, organophosphates, carbamates, and pyrethroids.

* Guthion® is a registered trademark of Mobay Chemical Corporation, Kansas City, Missouri.

** Lorsban® is a registered trademark of Dow Chemical Company, Midland, Michigan.

*** Thiodan® is a registered trademark of FMC Corporation, Philadelphia, Pennsylvania.

† Supracide® is a registered trademark of Ciba-Geigy Company, Basel, Switzerland.

‡ Zolone® is a registered trademark of Rhone-Poulenc, Monmouth Junction, New Jersey.

TABLE 1
Percentage of Selection Programs for Increased Pesticide Resistance in Predators and Parasitoids That Included Some Form of the Steps Listed Below

Phase of selection program[a]	Steps	Percentage		
		Predators[b]	Parasitoids[c]	Overall
I	Identify potentially effective natural enemies	100	100	100
	Determine trait(s) needing improvement	100	100	100
II	Adequate population sampling	38	45	42
	Production of genetic variability	56	50	56
	Efficient rearing and maintenance	100	100	100
	Selection program	100	100	100
III	Evaluation of natural enemy fitness	44	27	34
	Lab, greenhouse, or field cage evaluations of efficacy	19	9	13
	Genetic analysis of trait(s)	31	9	18
	Specific natural enemy/pest interaction models	0	5	3
IV	Determine field release strategy	13	9	11
	Release authorization	0	0	0
	Release and field efficacy evaluation	19	14	16
	Program implementation	6	0	3
	Cost analysis	6	0	3

[a]Phases defined as I = determination of needs; II = selection for resistance; III = postselection evaluation steps; and IV = release and establishment of selected strains.
[b]Number of predator/pesticide resistance genetic improvement projects assessed = 16.[10-17,19-21,42,43,46,60]
[c]Number of parasitoid/pesticide resistance genetic improvement projects assessed = 22.[22-25,27,30-41,45,47,48,52,61,62]

B. PHASE II: SELECTION FOR RESISTANCE

Steps involved in this phase of a selection program are adequate population sampling (Figure 1, step c), production of genetic variability (Figure 1, step d), efficient rearing and maintenance (Figure 1, step e), and selection for resistance (Figure 1, step f). Of the 38 selection programs examined, only 42% sampled more than three geographical locations for genetic differences in pesticide resistance (Table 1). Several studies did include some aspects of population sampling, but these commonly included less than ten populations. Exceptions to this trend include work by Roush and Hoy[15] (36 *M. occidentalis* populations) and Rosenheim and Hoy[33] (13 *Aphytis melinus* populations).

Genetic variation may be increased by pooling geographical strains of a natural enemy species, yet fewer than half of the selection programs used this approach (Table 1). Of the studies reviewed, none successfully used mutagens, radiation, or genetic engineering to increase the genetic variability within the selection programs surveyed. Havron et al.[34] used the mutagen ethyl methanesulfonate (EMS) to facilitate further selection within an *A. holoxanthus* colony previously selected for azinphosmethyl resistance. Exposure to the mutagen did not reduce fitness of the selected colony, but it did reduce azinphosmethyl resistance to levels below those earlier achieved through laboratory selection.

Most selection studies use effective rearing techniques, but the question remains as to what number of individuals should be used within single selection regimes. In genetic improvement of parasitoids, numbers used per selection round ranged from <200 individuals for *B. mellitor*[24] to >17,000 individuals for *A. lingnanensis*.[35] Of the parasitoid studies that reported the numbers of individuals used in selection regimes,[23,25,27,30-32,34-41] all used a minimum of 500 individuals during individual regimes, except those reported for *B. mellitor*,[24] and 44% used ≥1500 individuals. In contrast, individuals per selection regime in studies using predators ranged from approximately 115 for *C. carnea*[20,21] to 32,500 for *Amblyseius fallacis*.[13] However, only one used >500 individuals per selection round.[13] Survival per selection ranged from 30 to 55% in most studies.

Selection methods varied among species. For parasitoids, these methods included (1) direct application of insecticides through topical applications[24] or sprays,[27,36] (2) uptake of insecticides mixed with sucrose,[34,35,41] and (3) exposure to insecticide residues on treated surfaces.[22,23,25,30,31,39,40] Predators were exposed directly by either immersing[15,17] or spraying individuals[11,13,15,16,42] or by allowing predators to contact dried residues.[20,21] Parasitoids were always exposed to pesticides as adults. Only males of *Aphytis holoxanthus* and *A. lingnanensis* were exposed in studies by Javier et al.[41] Methods with the predators varied from selection of first instars in *C. carnea*[20,21] to all life stages in *Amblyseius fallacis*[13] and *P. persimilis*.[42] Some studies targeted the adult female predators only.[11,16] Overall, 22 and 68% of the selection programs using parasitoids[30,34,35,38,41] and predators,[11-17,21,43] respectively, successfully increased resistance to levels at which the natural enemy could survive applications of specific pesticides at recommended rates in the field as indicated by either laboratory or field assays.

C. PHASE III: POSTSELECTION EVALUATION

Steps involved in this phase involve evaluation of parasitoid fitness (Figure 1, step g); evaluation of efficacy in the laboratory, greenhouse, or field cages (Figure 1, step h); genetic analysis of resistance traits (Figure 1, step i); and development of specific natural enemy/pest interaction models (Figure 1, step j).

Increased levels of pesticide resistance may lead to reduced fitness.[44] Thus, it is important to assess any detrimental impact of laboratory selection on genetically improved organisms. Among all selection programs surveyed, 27 and 44% conducted fitness studies of selected parasitoids and predators, respectively (Table 1). Fitness was affected only in one parasitoid, *Aphytis lingnanensis*, which exhibited decreased fecundity and longevity after selection with azinphosmethyl.[41]

However, in five of seven studies on predators, fitness parameters (other than the ability to survive exposure to pesticides) were negatively impacted by selection for resistance. Resistance to carbaryl reduced survival of immatures and altered sex ratios (increased males) in *C. carnea*.[21] Selection for permethrin resistance in *M. occidentalis* resulted in genetic incompatibility for mating.[17] However, this may not be detrimental to long-term establishment of a polygenic trait. Higher mortality rates and reproductive disadvantages were found in strains of *P. persimilis* selected for pyrazophos and pirimiphos-methyl resistance.[10] In a sixth study, fitness was increased following selection for resistance to abamectin, which caused increased egg production in *M. occidentalis* females following exposure to the compound.[16]

Evaluation of the ability of a selected natural enemy to survive and effectively perform within an environment containing pesticide residues is essential for field implementation. Of the parasitoid and predator strains selected, 9 and 19%, respectively, underwent some type of residue evaluation prior to field release. These natural enemies included resistant strains of the parasitoids *A. melinus* [45] and *T. pallidus*[31,32,38] and the predators *M. occidentalis*,[46] *P. persimilis*,[43] and *C. carnea*.[21]

Mode of inheritance of resistance was assessed or generally defined in 9 and 31% of the parasitoid and predator selection programs, respectively. Studies on a carbaryl-resistant strain of *A. melinus* [47] and an azinphosmethyl-resistant strain of *T. pallidus*[48] indicated that mode of inheritance was neither fully dominant nor recessive, and in the latter species it was probably polygenic. Carbaryl resistance in a selected strain of *M. occidentalis* was found to be due to a single, nearly dominant gene.[46] A parathion-resistant strain of *P. persimilis* was found to be homozygous for a dominant factor which provided high resistance to parathion and lower levels of resistance to demeton-*S*-methyl.[42] Phosmet resistance in a selected strain of *Amblyseius nicholsi* was determined by a semidominant gene.[14] In contrast, mode of inheritance in a permethrin-resistant strain of *M. occidentalis* was not believed to be a simple recessive trait and was probably determined quantitatively.[17] Similarly, carbaryl resistance in *C. carnea* [21] was found to be due to a mechanism other than a single completely dominant or recessive gene.

Only one selection project (3% of total projects) included the development of a specific natural enemy/pest model to simulate actions of the improved natural enemy in the field (Table 1). This model was used to estimate the effects of biological and operational parameters on the introduction and establishment of an azinphosmethyl-resistant strain of the walnut aphid parasitoid *T. pallidus*.[49] This model provided insights into the development of a potentially effective release strategy.

D. PHASE IV: RELEASE AND ESTABLISHMENT OF SELECTED STRAINS

Steps in this phase involve determination of field release rates (Figure 1, step k), obtaining authorization to release (Figure 1, step l), release and field efficacy evaluations (Figure 1, step m), program implementation (Figure 1, step n), and cost analysis (Figure 1, step o). Only 6 of 38 ($\leq$16%) programs achieved this last phase (Table 1). Before releases, efforts were made in 11% of the selection programs to develop some type of release strategy to coordinate efforts to evaluate the results of the release. Exceptions to this occurred when selected strains were immediately exported to different locations as part of an introduction program.[41]

To date, no pesticide resistance selection program has been required to obtain authorization to release a genetically improved strain other than those permits that are normally required for the importation of wild-type individuals into a specific country or state. However, given increasing concerns over conservation of endemic species[50] and potential negative impacts from genetically altered arthropod strains,[51] specific release authorization at the national or state levels may be needed in the future.

Experimental field releases and evaluations have been conducted in 19 and 14% of the predator and parasitoid selection programs, respectively (Table 1). However, only the program with *M. occidentalis* has been fully implemented and examined with a cost/benefit analysis. Selected strains of *M. occidentalis* are commercially produced for mass releases in almond and apple orchards, resulting in substantial economic benefits to California agriculture.[19] Currently, efforts are underway to implement resistant strains of *Aphytis melinus*[30,52] and *T. pallidus*,[31,40] but results are still preliminary.[53]

III. IDENTIFICATION OF BOTTLENECKS IN THE PROCESS

Lack of adequate genetic variation for resistance may be an important bottleneck in the selection process. Few programs included extensive sampling to find resistance in field populations. Attempts to generate genetic variation by mutagenesis or other means were also rare. In selection programs that included field surveys, 25% (4/16) produced selected strains that were released for field tests as compared to 9% (2/22) of those that did not conduct preliminary field surveys. The percentage of programs that produced resistant strains released for field tests was 25 (5/20) for programs that attempted to increase genetic variation compared with 6 (1/18) for those that did not. Overall, 36% (4/11) of the studies that both conducted field surveys and increased genetic variability produced selected strains that were field released in contrast to 8% (2/27) of the studies in which only one or neither step was included. We cannot determine, however, if these steps caused greater success or were simply associated with other factors that contributed to success (e.g., more resources for a project).

IV. REALIZED HERITABILITY

Heritability is the proportion of phenotypic variance accounted for by additive genetic variation.[54] Heritability has been estimated for >1000 traits in animal populations.[55] Realized heritability (h^2) may be mathematically expressed as

$$h^2 = V_A(V_P)^{-1} \tag{1}$$

where V_A is the additive genotypic variance and V_P is the phenotypic variance.[54] Realized heritability can be estimated from selection studies as the response to selection (R) divided by the selection differential (S):

$$h^2 = R(S)^{-1} \tag{2}$$

TABLE 2
Estimates of Realized Heritability (h^2) and Number of Generations Required for a Tenfold Increase in Pesticide Resistance for Colonies of *Aphytis melinus* Using Responses to Selection (R) and Selection Differential (S)[a]

Compound	Colony	R	S	h^2	Generations
Carbaryl	Stutsman	0.047	0.174	0.268	21
	Aggregation	0.038	0.240	0.160	26
Chlorpyrifos	Moisi	0.015	0.184	0.080	68
	Aggregation	0.003	0.180	0.015	370
Dimethoate	Stutsman	0.029	0.300	0.098	34
	Aggregation	0.027	0.233	0.116	37
Malathion	Stutsman	0.045	0.165	0.274	22
	Aggregation	0.006	0.174	0.032	178
Methidathion	Pendery	0.045	0.207	0.219	22
	Aggregation	0.007	0.090	0.078	141

[a] Calculated from data from Rosenheim and Hoy.[30]

where $R = [\text{Log(final } LC_{50}) - \text{Log(initial } LC_{50})](\text{No. of generations selected})^{-1}$ and $S = i\,(s_P)$ where i is intensity of selection and s_P is the phenotypic standard deviation (see Falconer[54] and Tabashnik[56] for further details). Additionally, assuming that h^2 remains constant, the estimated number of generations (G) required for a natural enemy to develop tenfold resistance to a compound is R^{-1}.

Estimates of realized heritability provide a standardized summary of the results of selection experiments.[54,56] Under the same conditions required for probit analysis, this approach can be used to compare experiments which differ in selection intensity or duration. Estimation of heritability requires no assumptions about the mode of inheritance. However, like probit analysis, estimation of heritability is based on the assumption that the logarithm of tolerance to a pesticide is normally distributed.[57]

No one has reported heritability estimates for the numerous studies involving selection for pesticide resistance in natural enemies. We estimated realized heritability using selection data for the parasitoids *A. melinus*,[30] *A. holoxanthus*,[41] and *Ganaspidium utilis* Beardsley[58] (Table 2). Data necessary for heritability estimation were not available from most published studies on selection programs for predators.

A. *APHYTIS MELINUS*

For all pesticides and *A. melinus* colonies presented in Table 2, h^2 estimates varied from 0.015 to 0.274. We compared h^2 for Aggregation, a colony started from parasitoids collected at seven sites, and h^2 for colonies started from parasitoids collected at the single site that had the highest LC_{50} for each insecticide. The pooled parasitoid colony, Aggregation, had lower h^2 estimates than colonies started from single field sites in four of the five comparisons. In this case, pooling colonies did not increase genetic variation for resistance. However, the single sites were chosen because of their high tolerance, and thus, they might tend to have higher h^2 than the

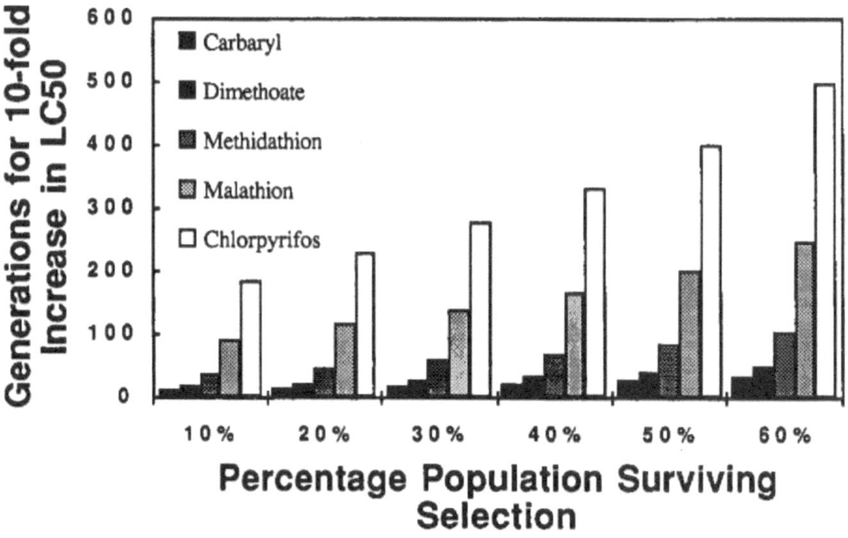

FIGURE 2. Estimated number of generations required for a tenfold increase in resistance in the aphelinid parasitoid *Aphytis melinus* DeBach (Aggregation colony; Rosenheim and Hoy[30]) as a function of selection intensity with various pesticides.

average population. If one did not know the relative tolerances of different populations, pooling might be more useful for increasing genetic variation. The estimated number of generations required for a tenfold increase in resistance for each *A. melinus* colony and pesticide ranged from 21 to 370 generations. Given these estimates and the level of pesticide resistance necessary for the selected strain to be effective in the field, minimum operational inputs and project lengths for specific selection projects could be predicted and decisions made as to the potential for success. Based on these estimates, one would be reluctant to continue selection for resistance to chlorpyrifos, malathion, and medidathion in the Aggregation colony if large increases (≥tenfold) in resistance were necessary. The number of generations required for a tenfold increase in resistance in the Aggregation colony to various pesticides under selection regimes ranging from 10 to 60% surviorship were estimated (Figure 2). This analysis suggests that selection for malathion and chlorpyrifos resistance using this parasitoid colony may be very time consuming and costly.

B. *APHYTIS HOLOXANTHUS*

Javier et al.[41] attempted to increase levels of malathion resistance in *A. holoxanthus* by selecting males only. Heritability estimates for the *A. holoxanthus* colony after 16, 20, 25, and 30 generations of selection were <0, which indicated that the colony did not respond to selection with malathion. This was the general conclusion drawn by the authors, but perhaps the project would have not been carried 30 generations if it had been known that the heritability estimate was –0.002 after 16 generations of selection.

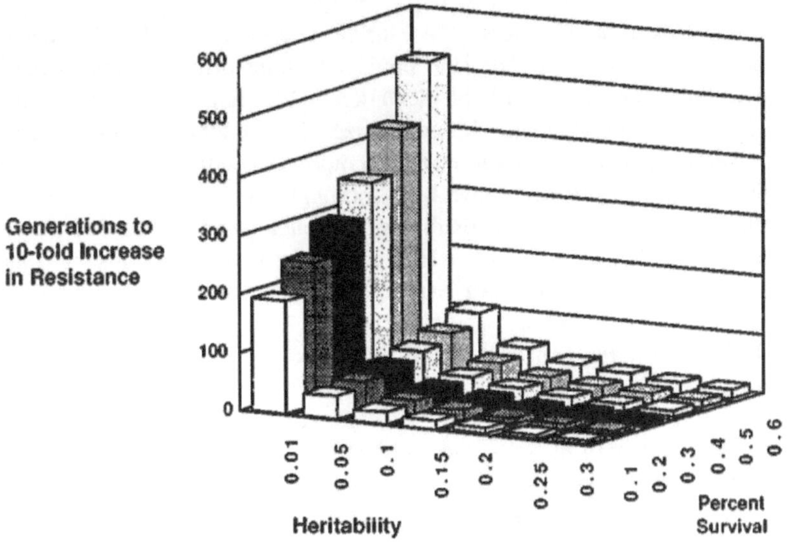

FIGURE 3. Effect of heritability and percent survival (of selected generations) on the number of generations required for a tenfold increase in resistance in hymenopterous parasitoids (based on a mean probit line slope of 3.38).

C. *GANASPIDIUM UTILIS*

We estimated heritability of resistance to fenvalerate in *G. utilis* Beardsley from data collected at the University of Hawaii, Honolulu.[58] Male and female parasitoids from a pooled colony (Select 1) had heritability values of 0.04 and 0.06, respectively, after 8 generations of selection at a mean percentage of survival of 23 in the selection regimes. We estimated that 25 generations would be required to increase fenvalerate resistance tenfold in females, with males requiring a longer time.

V. PREDICTING RESPONSE TO SELECTION

To give an overall perspective on the rates of response to selection, we calculated the number of generations needed to increase the LC_{50} tenfold for a range of heritability values and selection regimes (Figure 3) based on a mean slope of 3.38 for all selection studies assessed. Most of the estimated heritability values fell within the range of 0.01 to 0.25, and the mean percentages of survival in selection regimes were within 10 to 50%. These calculations suggest that selection programs in which the heritability of resistance is <0.05 may not be good prospects for continuation given the large number of generations required (>100) for a tenfold increase in LC_{50}.

VI. SUMMARY

Selection programs were more successful with predators than parasitoids with respect to field implementation. However, recent progress in parasitoid selection programs by Hoy at the University of Florida, Gainesville, and colleagues may change this trend. Overall, few projects produced strains that were implemented in the field. Reasons for the lack of implementation include limited time and resources, as well as lack of original intent. This emphasizes the point that selection, like other types of augmentation, is a solution that should be attempted after most other alternatives (e.g., conservation) have been thoroughly investigated. Analysis suggested that efforts to increase genetic diversity through identification of field populations exhibiting relatively high levels of resistance and the pooling of colonies collected from these populations was associated with greater success with respect to field implementation of strains.

The lack of a commonly used standardized index such as heritability to compare the relative abilities of various natural enemy strains to evolve resistance has probably hampered the efficiency of the selection process. Future incorporation of this estimate, like that of the LC_{50} and probit slopes, as an integral part of selection program progress analysis would bring a common denominator to all selection experiments and provide insights about genetic variation of selected natural enemies. Comparisons of heritability estimates for selections made on pure and aggregate strains of *A. melinus* suggest that pooling geographical strains did not increase heritability. Indeed, resistance developed faster in pure strains than in mixed strains. This observation contradicts the idea of maximizing genetic diversity by pooling geographic strains prior to initiation of selection. Instead, it may be best to survey field populations and use the most tolerant population as the source for individuals to be selected for resistance. Additional studies are needed to clarify this issue.

Advances in molecular biology may one day change our whole approach to typical laboratory selection regimes (i.e., exposure of individuals to pesticides with subsequent propagation of survivors).[7] It may be possible to obtain genes responsible for pesticide resistance in arthropod pests and insert them into natural enemies. Groundbreaking work toward this end has resulted in the insertion of genetic material from *Drosophila* spp./*Escherichia coli* into female *M. occidentalis*.[59]

ACKNOWLEDGMENTS

Manuscript reviews by Marjorie A. Hoy, Jay A. Rosenheim and K. M. Spollen were appreciated. This work was supported in part by U.S. Department of Agriculture grants 88-34135-3600 and 91-34135-6168. This paper is *Journal Series No. 3837* of the *Hawaii Institute of Tropical Agriculture and Human Resources.*

REFERENCES

1. **DeBach, P.,** The scope of biological control, in *Biological Control of Insect Pests and Weeds*, Reinhold, New York, 1964, 3.
2. **Mullin, C. A. and Croft, B. A.,** An update on development of selective pesticides favoring arthropod natural enemies, in *Biological Control of Agricultural Integrated Pest Management Systems*, Hoy, M. A. and Herzog, D. C., Eds., Academic Press, New York, 1985, 123.

3. **Croft, B. A. and Morse, J. G.,** Recent advances in natural enemy-pesticide research, *Entomophaga*, 24, 3, 1979.
4. **Hoy, M. A.,** Recent advances in genetics and genetic improvement of the Phytoseiidae, *Annu. Rev. Entomol.*, 30, 345, 1985.
5. **Beckendorf, S. K. and Hoy, M. A.,** Genetic improvement of arthropod natural enemies through selection, hybridization, or genetic engineering techniques, in *Biological Control in Agricultural IPM Systems*, Hoy, M. A. and Herzog, D. C., Eds., Academic Press, New York, 1985, 167.
6. **Croft, B. A.,** *Arthropod Biological Control Agents and Pesticides*, Wiley-Interscience, New York, 1990.
7. **Hoy, M. A.,** Genetic improvement of arthropod natural enemies: becoming a conventional tactic?, in *New Directions in Biological Control: Alternatives for Suppressing Agricultural Pests and Diseases*, Alan R. Liss, New York, 1990, 405.
8. **Hoy, M. A.,** Pesticide resistance in arthropod natural enemies: variability and selection responses, in *Pesticide Resistance in Arthropods*, Roush, R. T. and Tabashnik, B. E., Eds., Chapman and Hall, New York, 1990, 203.
9. **Hoy, M. A.,** Genetics and genetic improvement of the Phytoseiidae, in *Recent Advances in Knowledge of the Phytoseiidae*, Division of Agricultural Science, Publ. 3284, University of California, Berkeley, 1982, 72.
10. **Konig, K. and Hassan, S. A.,** Resistance and cross-resistance of the predacious mite *Phytoseiulus persimilis* (Athias-Henriot) to organophosphates, *J. Appl. Entomol.*, 101, 206, 1986.
11. **Markwick, N. P.,** Detecting variability and selecting for pesticide resistance in two species of phytoseiid mites, *Entomophaga*, 31, 225, 1986.
12. **Fournier, D., Pralavorio, M., Trottin-Caudal, Y., Coulon, J., Malezieux, S., and Berge, J. B.,** Selection artificielle pour la resistance au methidathion chez *Phytoseiulus persimilis* A. -H., *Entomophaga*, 32, 209, 1987.
13. **Strickler, K. and Croft, B. A.,** Selection for permethrin resistance in the predatory mite *Amblyseius fallacis*, *Entomol. Exp. Appl.*, 31, 339, 1982.
14. **Huang, M. D., Xiong, J. J., and Du, T. Y.,** The selection for and genetical analysis of phosmet resistance in *Amblyseius nicholsi*, *Acta Entomol. Sinica*, 30, 133, 1987.
15. **Roush, R. T. and Hoy, M. A.,** Genetic improvement of *Metaseiulus occidentalis*: selection with methomyl, dimethoate and carbaryl and genetic analysis of carbaryl resistance, *J. Econ. Entomol.*, 74, 138, 1981.
16. **Hoy, M. A. and Ouyang, Y. L.,** Selection of the western predatory mite, *Metaseiulus occidentalis* (Acari: Phytoseiidae), for resistance to abamectin, *J. Econ. Entomol.*, 82, 35, 1989.
17. **Hoy, M. A. and Knop, N. F.,** Selection for genetic analysis of permethrin resistance in *Metaseiulus occidentalis*: genetic improvement of a biological control agent, *Entomol. Exp. Appl.*, 30, 10, 1981.
18. **Hoy, M. A., Westigard, P. H., and Hoyt, S. C.,** Release and evaluation of a laboratory-selected strain of the predaceous mite *Metaseiulus occidentalis* (Acari: Phytoseiidae) in southern Oregon pear orchards and a Washington apple orchard, *J. Econ. Entomol.*, 76, 383, 1983.
19. **Headley, J. C. and Hoy, M. A.,** Benefit/cost analysis of an integrated mite management program for almonds, *J. Econ. Entomol.*, 80, 555, 1987.
20. **Grafton-Cardwell, E. E. and Hoy, M. A.,** Selection of the common green lacewing for resistance to carbaryl, *Calif. Agric.*, 40(9/10), 22, 1986.
21. **Grafton-Cardwell, E. E. and Hoy, M. A.,** Genetic improvement of common green lacewing, *Chrysoperla carnea* (Neuroptera: Chrysopidae): selection for carbaryl resistance, *Environ. Entomol.*, 15, 1130, 1986.
22. **Pielou, D. P. and Glasser, R. F.,** Selection for DDT resistance in a beneficial insect parasite, *Science*, 115, 117, 1952.
23. **Robertson, J. G.,** Changes in resistance to DDT in *Macrocentrus ancylivorus* Rohw. (Hymenoptera: Braconidae), *Can. J. Zool.*, 35, 629, 1957.
24. **Adams, C. H. and Cross, W. H.,** Insecticide resistance in *Bracon mellitor*, a parasite of the boll weevil, *J. Econ. Entomol.*, 60, 1016, 1967.
25. **Abdelrahman, I.,** Toxicity of malathion to the natural enemies of California red scale, *Aonidiella aurantii* (Mask.) (Homoptera: Diaspididae), *Aust. J. Agric. Res.*, 24, 119, 1973.

26. **Havron, A.,** Studies toward selection of *Aphytis* wasps for pesticide resistance, Ph.D. dissertation, Hebrew University, Jerusalem, Israel, 1983.
27. **Delorme, R., Angot, A., and Augé, D.,** Variations de sensibilité d'*Encarsia formosa* Gahan (Hym. Aphelinidae) soumis à des pressions de sélection insecticide: approches biologique et biochimique, *Agronomie*, 4, 305, 1984.
28. **Hsieh, C. Y.,** Effects of insecticides on *Diaeretiella rapae* (M'Intosh) with emphasis on bioassay techniques for aphid parasitoids, Ph.D. dissertation, University of California, Berkeley, 1984.
29. **Hoy, M. A.,** Developing insecticide resistance in insect and mite predators and opportunities for gene transfer, in *Biotechnology in Agricultural Chemistry*, American Chemical Society Series 334, American Chemical Society, Washington, D.C., 1987, 125.
30. **Rosenheim, J. A. and Hoy, M. A.,** Genetic improvement of a parasitoid biological control agent: artificial selection for insecticide resistance in *Aphytis melinus* (Hymenoptera: Aphelinidae), *J. Econ. Entomol.*, 81, 1539, 1988.
31. **Hoy, M. A. and Cave, F. E.,** Guthion-resistant strain of walnut aphid parasite, *Calif. Agric.*, 42, 4, 1988.
32. **Hoy, M. A., Cave, F. E., Beede, R. H., Grant, J., Krueger, W. H., Olson, W. H., Spollen, K. M., Barnett, W. W., and Hendricks, L. C.,** Release, dispersal, and recovery of a laboratory-selected strain of the walnut aphid parasite *Trioxys pallidus* (Hymenoptera: Aphidiidae) resistant to azinphosmethyl, *J. Econ. Entomol.*, 83, 89, 1990.
33. **Rosenheim, J. A. and Hoy, M. A.,** Intraspecific variation in levels of pesticide resistance in field populations of a parasitoid, *Aphytis melinus* (Hymenoptera: Aphelinidae): the role of past selection pressures, *J. Econ. Entomol.*, 79, 1161, 1986.
34. **Havron, A., Rosen, D., Prag, H., and Rossler, Y.,** Selection for pesticide resistance in *Aphytis*. I. *A. holoxanthus*, a parasite of the Florida red scale, *Entomol. Exp. Appl.*, 61, 221, 1991.
35. **Havron, A., Kenan, G., and Rosen, D.,** Selection for pesticide resistance in *Aphytis*. II. *A. lingnanensis*, a parasite of the California red scale, *Entomol. Exp. Appl.*, 61, 229, 1991.
36. **Delorme, R., Berthier, A., and Augé, D.,** The toxicity of two pyrethroids to *Encarsia formosa* and its host *Trialeurodes vaporariorum*: prospecting for a resistant strain of parasite, *Pestic. Sci.*, 16, 332, 1985.
37. **Havron, A., Rosen, D., and Rossler, Y.,** A test method for pesticide tolerance in minute parasitic Hymenoptera, *Entomophaga*, 32, 83, 1987.
38. **Hoy, M. A. and Cave, F. E.,** Toxicity of pesticides used on walnuts to a wild and azinphosmethyl-resistant strain of *Trioxys pallidus* (Hymenoptera: Aphidiidae), *J. Econ. Entomol.*, 82, 1585, 1989.
39. **Hoy, M. A. and Cave, F. E.,** Genetic improvement of a parasitoid: response of *Trioxys pallidus* to laboratory selection with azinphosmethyl, *Bio. Sci. Technol.*, 1, 31, 1991.
40. **Hoy, M. A., Cave, F. E., and Caprio, M. A.,** Guthion-resistant parasite ready for implementation in walnuts, *Calif. Agric.*, 45, 29, 1991.
41. **Javier, P. A., Havron, A., Morallo-Rejesus, B., and Rosen, D.,** Selection for pesticide resistance in *Aphytis*. III. Male selection, *Entomol. Exp. Appl.*, 61, 237, 1991.
42. **Schulten, G. G. M. and van de Klashorst, G.,** Genetics of resistance to parathion and demeton-S-methyl in *Phytoseiulus persimilis* A.-H. (Acari: Phytoseiidae), in *Proc. 4th Intern. Congr. Acarology*, Akademiai Kiado, Budapest, 1974, 519.
43. **Schulten, G. G. M., van de Klashorst, G., and Russell, V. M.,** Resistance of *Phytoseiulus persimilis* A.-H. (Acari: Phytoseiidae) to some insecticides, *Z. Angew. Entomol.*, 80, 337, 1976.
44. **Roush, R. T. and Daly, J. C.,** The role of population genetics in resistance research and management, in *Pesticide Resistance in Arthropods*, Roush, R. T. and Tabashnik, B. E., Eds., Chapman and Hall, New York, 1990, 97.
45. **Spollen, K. M. and Hoy, M. A.,** Residual toxicity of five citrus pesticides to a carbaryl-resistant and wild strain of the California red scale parasite *Aphytis melinus* DeBach (Hymenoptera: Aphelinidae), *J. Econ. Entomol.*, 86, 195, 1992.
46. **Roush, R. T. and Hoy, M. A.,** Laboratory, glasshouse, and field studies of artificially selected carbaryl resistance in *Metaseiulus occidentalis*, *J. Econ. Entomol.*, 74, 142, 1981.

47. **Spollen, K. M. and Hoy, M. A.,** Carbaryl resistance in a laboratory-selected strain of *Aphytis melinus* DeBach (Hymenoptera: Aphelinidae): mode of inheritance and implications for implementation in citrus IPM, *Biol. Control*, 2, 211, 1992.
48. **Brown, E. J., Cave, F. E., and Hoy, M. A.,** Mode of inheritance of azinphosmethyl resistance in a laboratory-selected strain of *Trioxys pallidus* Haliday. *Entomol. Exp. Appl.*, 63, 229, 1992.
49. **Caprio, M. A., Hoy, M. A., and Tabashnik, B. E.,** Model for implementing a genetically improved strain of a parasitoid, *Am. Entomol.*, 37, 232, 1991.
50. **Howarth, F. G.,** Environmental impacts of classical biological control, *Annu. Rev. Entomol.*, 36, 485, 1991.
51. **Hoy, M. A.,** Criteria for release of genetically-improved phytoseiids: an examination of the risks associated with release of biological control agents, *Exp. Appl. Acarol.*, 14, 393, 1992.
52. **Spollen, K. M. and Hoy, M. A.,** Resistance to carbaryl in populations of *Aphytis melinus* DeBach (Hymenoptera: Aphelinidae): implications for California citrus IPM, *Biol. Control*, 2, 203, 1992.
53. **Hoy, M. A.,** personal communication, 1992
54. **Falconer, D. S.,** *Introduction to Quantitative Genetics*, 3rd ed., Longman, New York, 1989.
55. **Mousseau, T. A. and Roff, D. A.,** Natural selection and and the heritability of fitness components, *Heredity*, 59, 181, 1987.
56. **Tabashnik, B. E.,** Resistance risk assessment: realized heritability of resistance to *Bacillus thuringiensis* in diamondback moth (Lepidoptera: Plutellidae), tobacco budworm (Lepidoptera: Noctuidae), and Colorado potato beetle (Coleoptera: Chrysomelidae), *J. Econ. Entomol.*, 85, 1551, 1992.
57. **Finney, D. J.,** *Probit Analysis*, 3rd ed., Cambridge University Press, Cambridge, 1971,
58. **Rathman, R. J.,** unpublished data, 1992
59. **Presnail, J. K. and Hoy, M. A.,** Stable genetic transformation of a beneficial arthropod, *Metaseiulus occidentalis* (Acari: Phytoseiidae) by a microinjection technique, *Proc. Natl. Acad. Sci. U.S.A.*, 89, 7732, 1992.
60. **Field, R. P. and Hoy, M. A.,** Evaluation of genetically improved strains of *Metaseiulus occidentalis* (Nesbitt) (Acarina: Phytoseiidae) for integrated control of spider mites in greenhouses, *Hilgardia*, 54, 1, 1986.
61. **Rosenheim, J. A. and Hoy, M. A.,** Sublethal effects of pesticides on the parasitoid *Aphytis melinus* (Hymenoptera: Aphelinidae), *J. Econ. Entomol.*, 81, 476, 1988.
62. **Spollen, K. M. and Hoy, M. A.,** Genetic improvement of an arthropod natural enemy: relative fitness of a carbaryl-resistant strain of the California red scale parasite *Aphytis melinus* DeBach, *Biol. Control*, 2, 87, 1992.

Chapter 6

GENETIC ANALYSIS OF THE RESISTANCE OF TWO STRAINS OF *TYPHLODROMUS PYRI* TO SYNTHETIC PYRETHROID INSECTICIDES

N. P. Markwick

TABLE OF CONTENTS

0-8493-2607-9/94/$0.00+$.50

I. INTRODUCTION

Pyrethroid insecticides (SPs) are very effective against many lepidopterous pests of horticultural crops. In New Zealand, the major pests of pipfruit are codling moth *Cydia pomonella* L. and light brown apple moth (LBAM) *Epiphyas postvittana* (Walker). Other lepidopterans which cause damage to pipfruit are the native leafrollers (*Ctenopseustis* spp. and *Planotortrix* spp.). Over the past 30 years in New Zealand, these pests have been controlled with organophosphate insecticides (OPs). However, SPs provide an opportunity for the orchardist to control these pests with products less hazardous during application, used at lower rates, and with fewer sprays per season.[1]

This opportunity has been prevented until recently because pyrethroids have one serious drawback; they have been associated with severe outbreaks of tetranychid mites.[2] Generally, this can be attributed to the destruction of natural enemies, especially predatory phytoseiid mites. However, in New Zealand, field and laboratory selection programs have now produced two strains of the predatory mite *Typhlodromus pyri* Scheuten, which are SP resistant.[2-4] Field trials over the past four seasons have shown that control of European red mite (*Panonychus ulmi* Koch) by these SP-resistant (SP-R) predators under an SP-spray program is very good. These strains are also OP resistant.[1,4] Hence, the use of pyrethroids in pipfruit orchards where SP-R predatory mites have been released is now a viable alternative to the traditional OP program.[1]

This chapter describes a complementary laboratory program investigating aspects of the genetics of the resistance. The main objectives of this genetic analysis were to determine whether the SP resistance (1) showed maternal effects, (2) was dominant, (3) was single gene or polygenic, and (4) was inherited in the same way in both strains of SP-R *T. pyri* (field and laboratory selected). The answers to these questions should assist in the successful establishment and management of the SP-R predators in integrated mite control programs in New Zealand pipfruit orchards.

II. METHODS

A. CONCENTRATION/MORTALITY TESTS OF PARENTAL STRAINS

Before a genetic analysis of the SP-resistant strains of *T. pyri* from the field and laboratory was undertaken, it was essential to determine the concentration/mortality responses of susceptible and resistant strains. These were estimated from a series of tests using the method described in Markwick[3] in which mature female predators (20 per replicate) were sprayed on bean leaf disks under a Potter Tower. At least four (usually more) concentrations of cypermethrin (Cymbush 6.5 DG, ICI) were used for each test. Resistance factors (RF = LC_{50} resistant strain/LC_{50} susceptible strain) were also calculated.

B. SOURCES OF MITES

SP-susceptible (SP-S) *T. pyri* were collected in the Havelock North area of Hawkes Bay from orchards which had never been exposed to SP sprays or even to the possibility of SP-spray drift.

SP-R predators (field strain) were collected from the block of apple trees (cv. Splendour) at the DSIR Research Orchard at Appleby (Nelson) where this resistant strain was produced by field selection.[4]

SP-R predators (laboratory strain) were taken from the laboratory-selected strain in culture at DSIR and also collected from SP-sprayed apple trees at the DSIR Kumeu Research Orchard (Auckland) where the laboratory-selected strain had been released previously.

One source of variation in estimating LC_{50} values was found to be the level of feeding by the predators just prior to the test.[5] *T. pyri* from Havelock North were well-fed, as were those from the laboratory culture. However, the predators from Appleby were not as well-fed because the European red mite population was low and well under control by the predators. Thus, to avoid such variation, these *T. pyri* were fed in the laboratory before the concentration/mortality tests were done. This was necessary to give valid comparisons with the SP-S Havelock North and SP-R laboratory strains and with the well-fed F_1, F_2, and backcross progeny from the ensuing crosses.

C. CROSSING EXPERIMENTS

Following the estimation of concentration/mortality curves (c/m curves) and LC_{50} values for female predators of one strain of SP-S and two strains of SP-R *T. pyri*, crosses and reciprocal crosses between resistant and susceptible strains were carried out (Table 1).

While F_1 progeny of susceptible and resistant parents give information about the dominance of the inherited character and possible maternal effects, F_2 and backcross generations indicate whether the inheritance is due to a single gene or is polygenic. F_2 progeny were obtained by crossing F_1 progeny from each of the crosses and reciprocal crosses. Backcross progeny were obtained by crossing virgin F_1 females (from SP-R × SP-S cross) with SP-S males. The reciprocal cross (SP-S female × F_1 male) is not possible because males are haploid in this species, and thus, heterozygote males do not exist (see Sections III and IV).

Crosses and reciprocal crosses were also made between females from one resistant strain and males from the other resistant strain (Table 1) in order to determine whether the similar phenotypes in the two SP-R strains of *T. pyri* are controlled by similar genotypes. If they are, parents and their F_1 progeny share the same concentration/mortality curve.

The virginity of female predators, prior to each cross, was ensured by dividing the acrylic rearing plates[3] into ten small squares with Stickem®* and by placing one nymph (either proto- or deutonymph) in each square. When it had moulted to an adult, it was transferred to a rearing plate set up in the usual way with pollen, honey and water solution, and a two-spotted mite. Here, mating between males and females of the appropriate strains could take place.

A minimum of 25 virgin females was used in each cross. The number of males per cross was variable, but lower, since the ratio of males in the population is quite low.

*Stickem® is a registered trademark of Davis Gelatin Ltd., Christchurch, New Zealand.

TABLE 1
Summary of Genetic Crosses Performed with Two Synthetic Pyrethroid-Resistant (SP-R) Strains and One SP-Susceptible (SP-S) Strain of *T. pyri*

Crosses between SP-R and SP-S strains

Parental Crosses	(F_1 Progeny)	Reciprocal Crosses	(F_1 Progeny)
Lab[a] ♀ × Sus[b] ♂	(LS♀ + L♂)	Sus ♀ × Lab ♂	(SL♀ + S♂)
(70) (22)		(90) (37)	
FLD[c] ♀ × Sus ♂	(FS♀ + F♂)	Sus ♀ × FLD ♂	(SF♀ + S♂)
(36) (10)		(70) (21)	

F_1 Crosses	(F_2 Progeny)	Reciprocal Crosses	(F_2 Progeny)
LS ♀ × L ♂	(LL + LS)	SL ♀ × S ♂	(SL + SS)
(96) (19)		(65) (13)	
FS ♀ × F ♀	(FF + FS)	SF ♀ × S ♂	(SF + SS)
(60) (15)		(58) (9)	

Backcrosses	(backcross ♀ progeny)
F_1♀(LS) × S ♂	(SL + SS)$^+$
(121) (24)	
F_1♀(FS) × S ♂	(FS + SS)$^+$
(35) (8)	

Crosses between two SP-R strains

Parental crosses	(F_1 ♀ progeny)
FLD ♀ × Lab ♂	(FL)
(99) (47)	
Lab ♀ × FLD ♂	(LF)
(102) (46)	

[a] Lab (L) = laboratory-selected SP-R strain.
[b] Sus (S) = SP-S strain from the Havelock North area of Hawkes Bay.
[c] FLD (F) = field-selected SP-R strain from Appleby, Nelson.

The ♀ to ♂ ratio is approximately 60:40 at egg hatch[6] and 2:1 in the field.[7] Parents were transferred to new rearing plates every 6 to 7 days to ensure that no mixing of the generations occurred.

All rearing and crossing experiments were carried out in the laboratory on acrylic plates, and numbers were sometimes inadequate to give good fitting concentration/mortality curves.

D. STATISTICAL ANALYSIS

Concentration/mortality curves were derived for the parental strains and the progeny of the crosses. LC_{50} values (and 95% confidence limits) were estimated using the maximum likelihood program (MLP).[8] This program was also used to determine whether the difference in concentration/mortality curves were statistically significant by comparing the position and parallelism of the curves with χ^2 goodness-of-fit tests.

TABLE 2
LC_{50} Values, 95% Confidence Limits, Heterogeneity Factors, and Numbers of Female *Typhlodromus pyri* Tested for Each Parental Strain and Their F_1 Progeny

Strain/cross	Numbers tested	LC_{50} a.i.%	95% confidence limits	Heterogeneity factor
Sus (S)[a]	152	0.0004	0.00003–0.0006	0.82
Lab (L)[b]	437	0.0115	0.0087–0.0153	0.14
FLD (F)[c]	465	0.0110	0.0080–0.0143	1.63
F_1 (LS)	163	0.0047	0.0025–0.0078	1.10
F_1 (SL)	246	0.0036	0.0019–0.0052	1.85
F_1 (LS+SL)	409	0.0039	0.0024–0.0056	2.21
F_1 (FS)	109	0.0029	0.0017–0.0043	0.38
F_1 (SF)	265	0.0039	0.0020–0.0059	2.20
F_1 (FS+SF)	374	0.0036	0.0028–0.0045	1.86

[a] Sus (S) = SP-S strain from the Havelock North area of Hawkes Bay.
[b] Lab (L) = laboratory-selected SP-R strain.
[c] FLD (F) = field-selected SP-R strain from Appleby, Nelson.

III. RESULTS

The mode of sex determination of males is not known for *T. pyri*. It is known that female *T. pyri* are diploid and males are haploid (2n = 8, n = 4),[9] but it has not been determined whether the haploid males arise through unfertilized eggs as in arrhenotoky or whether they have a biparental origin as do the males of three other phytoseiid species.[10,11] The latter seems likely, since unmated female *T. pyri* do not oviposit. However, by whichever method the haploid males originate, they inherit all their genetic material from their mothers. This must be kept in mind when determining the genotypes of backcross and F_2 generations.

Table 2 gives estimated LC_{50} values, 95% confidence limits, heterogeneity factors, and the numbers of females tested for each parental strain and their F_1 progeny.

A. MATERNAL EFFECTS

Concentration/mortality curves for F_1 crosses and reciprocal crosses were compared by χ^2 analysis for position and parallelism. No significant differences were found for F_1 females from crosses and reciprocal crosses between the SP-S and SP-R (laboratory) strains (p >0.10 position, p >0.5 parallelism) or for F_1 females from SP-S and SP-R (field) crosses and reciprocal crosses (p >0.1 position, p >0.9 parallelism). This indicates that there are no maternal effects in the inheritance of the SP-resistance in either SP-R strain of *T. pyri*. Thus, the F_1 data from the concentration/mortality tests for these crosses and reciprocal crosses were combined to give a single F_1 curve (Figures 1 and 2).

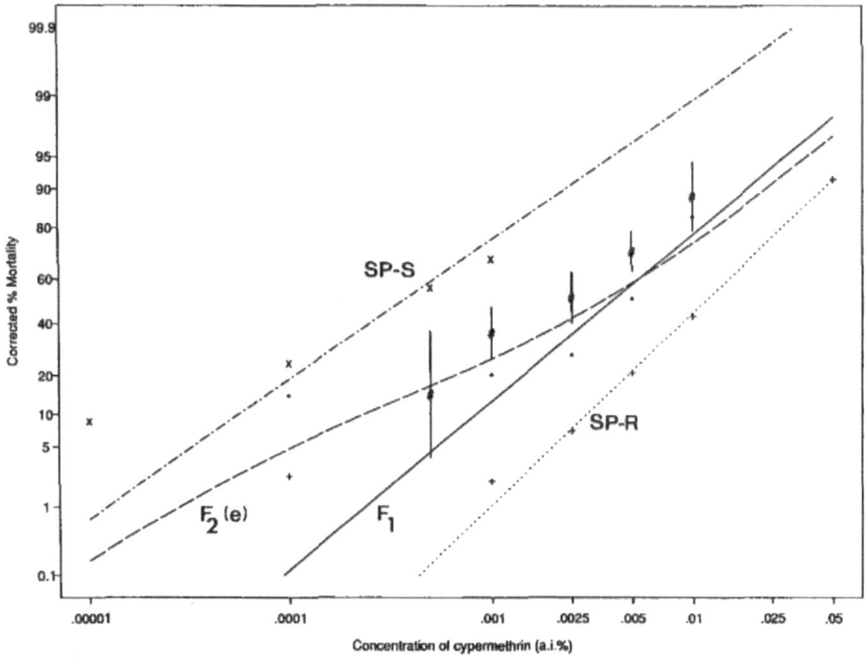

FIGURE 1. Concentration/mortality curves for an SP-susceptible strain (._._.) and a laboratory-selected SP-resistant strain of *T. pyri* (.....) and the F_1 progeny from reciprocal crosses between them (——). The expected mortality curve (-----) and observed mortalities (#) for F_2 female progeny are also shown.

B. DOMINANCE

The F_1 curves lie between those of the parents and are displaced toward that of the resistant parent (Figures 1 and 2). Stone[12] gave a mathematical formula (from Falconer[13]) which he felt defined more precisely all levels of dominance.

$$D = \frac{2X_2 - X_1 - X_3}{X_1 - X_3}$$

where $X_1 = \log_{10}(LC_{50})$ of the SP-R strain, $X_2 = \log_{10}(LC_{50})$ of the F_1 females, $X_3 = \log_{10}(LC_{50})$ of the SP-S strain, and $1 > D > 0$ indicates incomplete dominance.

For F_1 progeny of the SP-S and SP-R (laboratory) strains, $D = 0.3823$, and for F_1 progeny of SP-S and SP-R (field) strains, $D = 0.3485$. Thus, both SP-resistant strains of *T. pyri* exhibit incomplete dominance for SP resistance.

This description of the SP resistance is independent of the number of genes involved in resistance determination and would be the same whether it was due to a single gene or was polygenic.

C. SINGLE OR POLYGENIC RESISTANCE DETERMINATION

Results from testing F_2 and backcross progeny can indicate whether the SP resistance is determined by a single gene or polygenically. LC_{50} values, 95% confi-

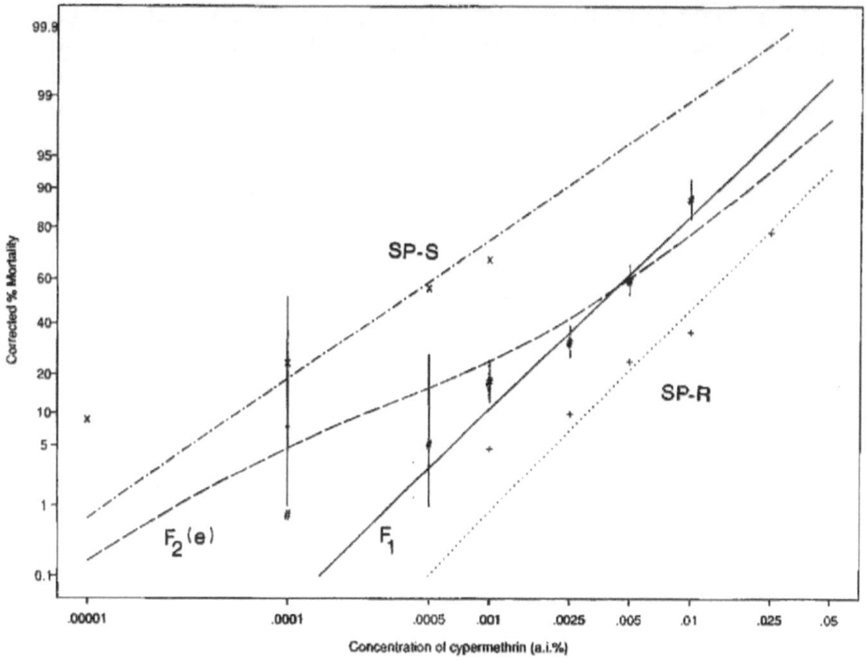

FIGURE 2. Concentration/mortality curves for an SP-susceptible strain (- - -) and a field-selected SP-resistant strain of *T. pyri* (.....) and the F_1 progeny from reciprocal crosses between them (———). The expected mortality curve (-----) and observed mortalities (#) for F_2 female progeny are also shown.

dence limits, and heterogeneity factors calculated using MLP and the numbers tested for F_2 and backcross progeny are given in Table 3.

If the inheritance of the SP resistance is through a single gene, then the expected mortalities for F_2 females at each concentration can be calculated by the following formula (from Georghiou[14]):

$$X_y = W_{(SS)}0.25 + W_{(SR)}0.50 + W_{(RR)}0.25$$

For *T. pyri*, this is only the case if the crosses and reciprocal crosses are combined, since males are haploid and inherit all their genetic material from their mothers. If the F_1 male is derived from a susceptible female X resistant male cross, it will have only SP-S genes. The F_2 female progeny resulting from a cross between this F_1 male and an F_1 female will have only SS and SR genotypes. This is the same as for the backcross between an F_1 female and a susceptible male. In either of these cases, the expected mortalities can be calculated by the following formula:

$$X_y = W_{(SS)}0.50 + W_{(SR)}0.50$$

The reciprocal cross means that F_2 female progeny derived from an F_1 female and an F_1 male with only SP-R genes will have only RR and RS genotypes, the same as

TABLE 3
LC_{50} Values, 95% Confidence Limits, Heterogeneity Factors, and Numbers of Female *Typhlodromus pyri* Tested for F_2 and Backcross (bc) Progeny

Progeny/ cross	Numbers tested	LC_{50} a.i.%	95% confidence limits	Heterogenity factor
F_2, LS/LS	142	0.0025	0.0013–0.0041	0.37
F_2, SL/SL	96	0.0014	0.0001–0.0025	0.96
F_2, combined	238	0.0020	0.0012–0.0028	0.46
F_2, FS/FS	95	0.0020	0.0001–0.0044	0.01
F_2, SF/SF	137	0.0023	0.0015–0.0032	1.20
F_2, combined	232	0.0022	0.0015–0.0030	0.51
bc, LS/S	224	0.0024	0.0001–0.0077	3.07
bc, FS/S	114	0.0029	0.0010–0.0044	0.58
F_2 + bc, LS	320	0.0019	0.0006–0.0035	2.77
F_2 + bc, FS	251	0.0025	0.0017–0.0033	1.11

for the backcross between an F_1 female and an SP-R male. For female progeny from either of these crosses, the formula for calculating expected mortalities is

$$X_y = W_{(RS)}0.50 + W_{(RR)}0.50$$

The *T. pyri* data were applied to these formulas and expected and observed mortalities were first tested using the standard χ^2 method derived from Finney,[15]

$$\chi^2 = (O - np)^2 / [np(1 - p)]$$

where O is the observed number dead, n is the total number of mites treated, and p is the expected proportion affected (dead).

However, Preisler, Hoy, and Robertson[16] suggested that the use of the standard χ^2 statistic does not describe the data well in many cases because it assumes that the only source of variability is the random binomial variation. Other sources of variation ("overdispersion") in experiments used to determine mode of inheritance include slight differences in temperature and humidity over time or in dose preparation between replicates. Further, the standard χ^2 statistic assumes that sample sizes are large. Outliers and between experiment variations can also be a source of extra variation in experiments where all concentration/mortality bioassays were not done simultaneously. This was certainly the case with the present study. Thus, it seemed appropriate to use their modified χ^2 statistic which was developed for the analysis of bioassay data with extra sources of variation.

A χ^2 goodness-of-fit statistic was calculated for each of the parental curves, F_1, F_2, and backcross (see Preisler, Hoy, and Robertson, Equation 4).[16] The heterogeneity factors resulting from these calculations (χ^2/d.f.) were then pooled separately for the laboratory and field SP-R strains. The variance, vâr (O) in the denominator of χ^2

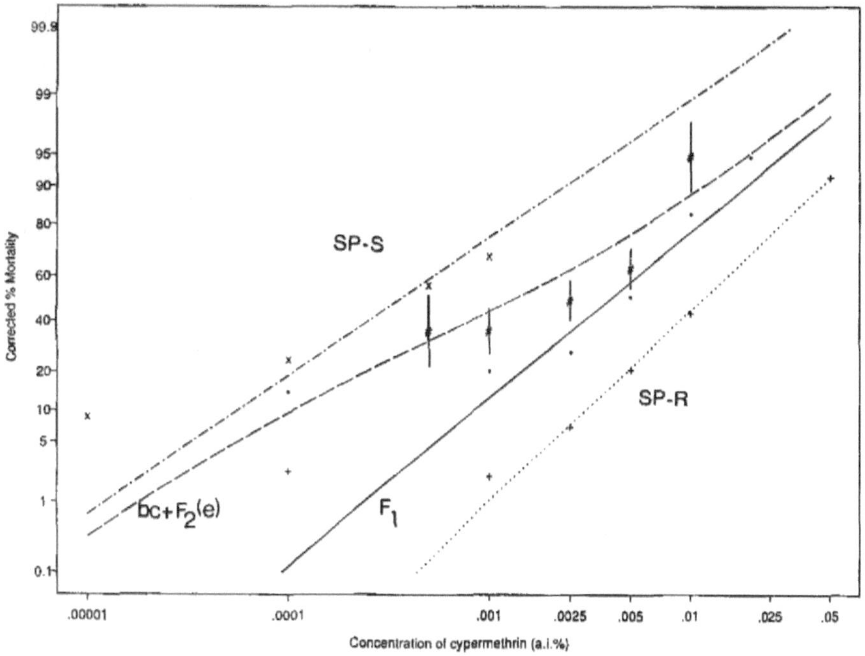

FIGURE 3. The expected concentration/mortality curve and observed mortalities (#) for backcross plus F_2 female progeny for SP-S and SP-R (laboratory) *T. pyri* (-----). F_1 (——) is combined for reciprocal crosses. The backcross is between F_1♀ (laboratory × SP-S) and SP-S♂, and F_2 is the progeny of the F_1 cross (SP-S♀ × SP-R laboratory ♂).

statistic below, for testing the single model (Preisler, Hoy, and Robertson, Equation 5)[16] was then multiplied by the pooled heterogeneity estimate.

$$\chi^2 = (O - np)^2 / \hat{var}(O)$$

1. Laboratory-Selected SP-R *T. pyri*

When observed and expected mortalities of the combined F_2 female progeny were compared by the above method, they were not significantly different ($\chi^2 = 6.53$ on 5d.f., $p > 0.05$). The expected F_2 concentration/mortality curve and observed mortalities are shown in Figure 1.

Observed and expected mortalities of females from the backcross (F_1♀ × SP-S♂) combined with those from the F_2 (F_1♀ × F_1♂ with only SP-S genes) generation were also not significantly different ($\chi^2 = 8.98$ on 5d.f., $p > 0.05$). The expected concentration/mortality curve for F_2 plus backcross females and the observed mortalities are shown in Figure 3.

This data is consistent with the single-gene model.

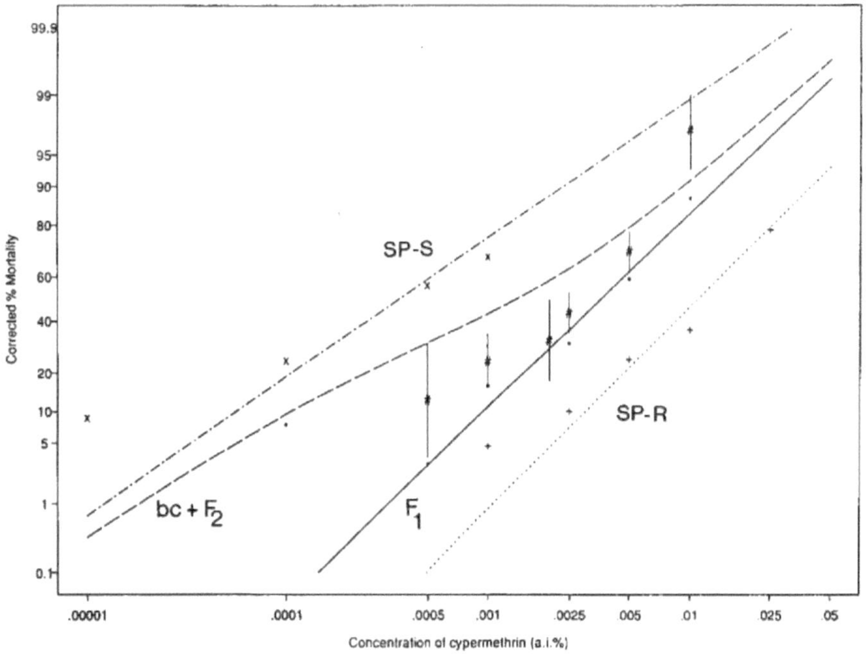

FIGURE 4. The expected concentration/mortality curve and observed mortalities (#) for backcross plus F_2 female progeny for SP-S and SP-R (field) *T. pyri* (-----). F_1(——) is combined for reciprocal crosses. The backcross is between F_1♀ (field × SP-S) and SP-S♂, and F_2 is the progeny of the F_1 cross (SP-S♀ × SP-R field ♂).

2. Field-Selected SP-R *T. pyri*

When observed and expected mortalities of the combined F_2 progeny were compared by the above method, they were not significantly different ($\chi^2 = 8.5$ on 6d.f., $p > 0.05$). Figure 2 shows the expected F_2 concentration/mortality curve and the observed mortalities.

However, observed mortalities for backcross females were significantly lower than expected at the lower concentrations of cypermethrin, giving a significant overall χ^2 value of 14.84 on 6d.f. ($p < 0.05$). When observed mortalities of females from the backcross (F_1♀ X SP-S♂) were combined with those from the F_2 (F_1♀ X F_1♂ with only SP-S genes) generation, they were also significantly different from the expected values ($p < 0.01$). Figure 4 shows concentration/mortality curves for F_1 and combined F_2 plus backcross female progeny.

Thus, the results for F_2 progeny of the field-selected SP-R *T. pyri* suggest that the resistance is due to a single gene, but this is not supported by the backcross results for this strain.

Despite the backcross results for the field-selected strain, analysis of the other three crosses suggest that the SP resistance in both SP-R strains of *T. pyri* follows the single-gene model, although it is recognized that the results are not conclusive.

TABLE 4
LC_{50} Values, 95% Confidence Limits, Numbers Tested, and Heterogeneity Factors for F_1 Progeny of the Two SP-R Strains of *Typhlodromus pyri*; Results of D-M Tests for the Two Parental SP-R Strains (as in Table 2) Are Also Given

Strain/ cross	Numbers tested	LC_{50} a.i.%	95% confidence limits	Heterogeneity factor
Lab (L)[a]	437	0.0115	0.0087–0.0153	0.14
FLD (F)[b]	465	0.0110	0.0080–0.0143	1.63
F_1 LF	254	0.0084	0.0015–0.0133	2.93
F_1 FL	169	0.0077	0.0048–0.0113	0.43

[a] Lab (L) = laboratory-selected SP-R strain.
[b] FLD (F) = field-selected SP-R strain from Appleby, Nelson.

D. NATURE OF SP-RESISTANCE IN THE TWO SP-R STRAINS

Crosses and reciprocal crosses between SP-R *T. pyri* from the field- and laboratory-selected strains were undertaken (Table 4) and then compared with the χ^2 test using MLP. Concentration/mortality curves for F_1 female progeny from these crosses and reciprocal crosses (Figure 5) were not significantly different from each other nor from either of the SP-R parents (position $p > 0.05$; parallelism $p > 0.25$). This indicates that F_1 and parental females from the two SP-R strains share the same concentration/mortality curves and thus a similar genetic mechanism.

IV. DISCUSSION

The results clearly indicate that the SP-resistance developed in the field- and laboratory-selected strains of *T. pyri* does not show maternal effects and is incompletely dominant. The resistance is probably controlled by a similar genetic mechanism in both strains.

It is not so clear whether the resistance is polygenic or controlled by a single gene. The inconclusive results probably reflect the relatively small numbers of females available for the concentration/mortality tests. However, application of the formulas derived by Georghiou[14] from the theory of single factor Mendelian inheritance to these results combined with the use of a modified χ^2 statistic[16] generally supports a single-gene model. In three of the four crosses involving F_2 and backcross progeny, observed and expected mortalities were not significantly different.

In the fourth of these crosses, the lower than expected mortality of backcross female progeny from the field strain may indicate that in this strain an additional minor gene (or genes) is influencing the resistance which is determined by a single incompletely dominant gene. Although this is speculative and more work needs to be done, previous examples of this phenomenon occur in the literature.[17-19] In these three cases, the inheritance of insect resistance to synthetic pyrethroids was being investigated and in all three, lower than expected mortalities occurred in the backcross. All

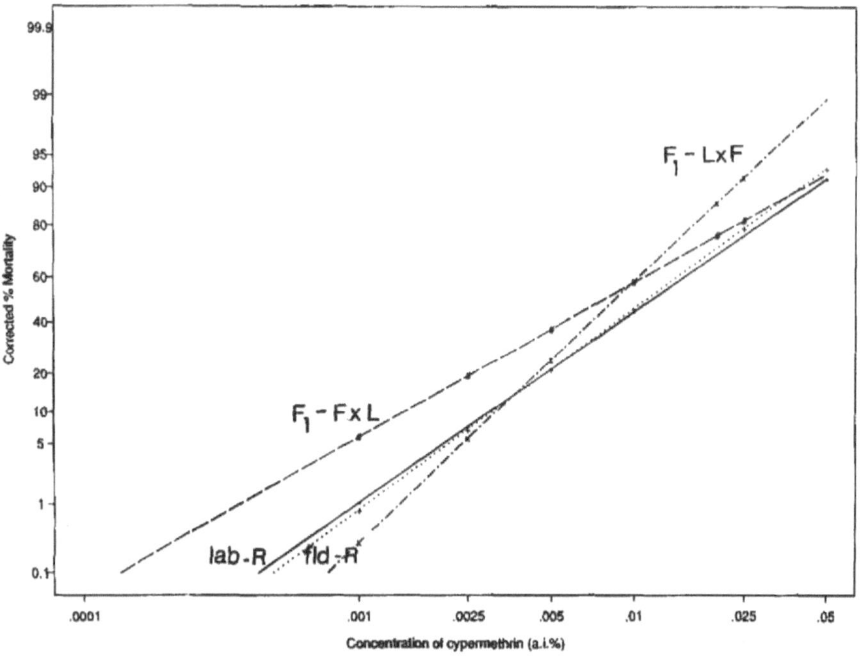

FIGURE 5. Concentration/mortality curves for parental laboratory (——) and field (.....) *SP-R T. pyri* and for F_1 progeny of reciprocal crosses and between these two SP-resistant strains: laboratory ♀ × field ♂ (_._._) and field ♀ × laboratory ♂ (-----).

of the authors involved suggested that, while the resistance was inherited through a single major gene, it was influenced by additional minor resistance factors.

That the pyrethroid resistance in these *T. pyri* strains is inherited through a single gene seems probable, moreover, because the two strains appear to have a similar genetic mechanism. The possibility that two strains, selected independently, would develop a similar polygenic mechanism, even when some of them had been derived from the same source, seems remote. In the present study, the laboratory-selected strain probably benefitted from the preselection in the field due to spray drift.[3]

Selective breeding programs have produced SP resistance in other phytoseiid mites. Generally, laboratory selection has resulted in a resistance which is polygenically determined. This was true of the permethrin resistance developed in *Metaseiulus occidentalis* by Hoy and Knop[20] in California and in *Amblyseius fallacis*[21,22] in Michigan. However, in Australia, a strain of *M. occidentalis* selected with bioresmethrin developed an SP resistance which appeared to be inherited through a dominant system, with the involvement of more than one factor.[23] It was also found that when these factors were passed on through the female parent they appeared to convey a greater resistance than when passed on by the male. More recently, a strain of *Phytoseiulus persimilis* was reported to have increased its resistance to permethrin by 18-fold,[24] and a strain of *M. occidentalis* was reported to have developed resistance to permethrin.[25] The mode of inheritance data for this strain suggested that the resistance is partially dominant and controlled by a single major gene.[26]

Because the question of single-gene or polygenic inheritance is not conclusively resolved in these SP-resistant strains of *T. pyri*, it is suggested that orchard management practices should include spraying any SP-predator release sites with an SP insecticide prior to inoculation in order to eliminate susceptible predators. In New Zealand, where pyrethroid insecticides will probably be applied to pipfruit orchards three to five times per season, no difficulty in retaining SP resistance is anticipated. Field trials to date have supported this.[1] Where SP insecticides are applied only once early in each season, the situation is less clear. However, evaluation trials have begun, and after two seasons of such treatment at Havelock North Research Orchard, SP resistance in *T. pyri* had been retained at a useful level. Continuing trials involving reduced pyrethroid spraying during the season will test the sustainability of the SP resistance in *T. pyri*.

ACKNOWLEDGMENTS

I wish to thank Vanessa White and Peter Shaw for field collections of *T. pyri* from Havelock North and Appleby, respectively, and John McKenzie at the University of Melbourne for his helpful interpretation of the earlier data and suggestions for further testing to improve the analysis. I would also like to thank Howard Wearing, HortResearch, Clyde and Marjorie Hoy, University of Florida, Gainesville for their assistance with the manuscript. John Maindonald and Roderick Ball, HortResearch, have given invaluable help with the MLP program.

REFERENCES

1. **Walker, J. T. S., Markwick N. P., Wearing C. H., Shaw P. W., and White V.,** Pyrethroid insecticides for apple pest control. II. Field evaluation of mite and insect control, *Proc. N.Z. Weed Pest Control Conf.*, 43, 301, 1990.
2. **Markwick, N. P.,** Induced resistance in beneficials with particular reference to *Typhlodromus pyri* Scheuten (Acarina: Phytoseiidae), in *Proc. Joint* DSIR/CSIRO *Workshop*, Daly, J. C. and Suckling, D. M., Eds., CSIRO Site Services, Canberra, Australia, 1986, 91.
3. **Markwick, N. P.,** Detecting variability and selecting for pesticide resistance in two species of phytoseiid mites, *Entomophaga*, 31, 225, 1986.
4. **Markwick, N. P., Wearing, C. H., and Shaw, P. W.,** Pyrethroid insecticides for apple pest control. I. Development of pyrethroid-resistant predatory mites, *Proc. N.Z. Weed Pest Control Conf.*, 43, 296, 1990.
5. **Markwick, N. P.,** unpublished data, 1988.
6. **Hayes, A. J.,** Simulation models on the European red mite *Panonychus ulmi* and its predator *Typhlodromus pyri* in New Zealand apple orchards, Ph.D. thesis, University of Auckland, New Zealand, 1986, 228.

7. **Chant, D. A.,** Phytoseiid mites (Acarina: Phytoseiidae). I. Bionomics of seven species in southeastern England, *Can. Entomol.*, 91(Suppl. 12), 166, 1959.
8. **Ross, G. J. S.,** MLP Maximum Likelihood Program, Rothamsted Experimental Station, Harpenden, U.K., 1980.
9. **Wysoki, M. and Bolland, H. R.,** Chromosome studies of phytoseiid mites (Acari: Gamasida), *Int. J. Acarol.*, 9, 91, 1983.
10. **Helle, W., Bolland, H. R., van Arendonk, R., de Boer, R., Schulten, G. G. M., and Russell, V. M.,** Genetic evidence for biparental males in haplo-diploid predator mites (Acarina: Phytoseiidae), *Genetica*, 48, 165, 1978.
11. **Hoy, M. A.,** Parahaploidy of the "arrhenotokous" predator, *Metaseiulus occidentalis* (Acarina: Phytoseiidae) demonstrated by X-irradiation of males, *Entomol. Exp. Appl.*, 26, 97, 1979.
12. **Stone, B. F.,** A formula for determining degree of dominance in cases of monofactorial inheritance of resistance to chemicals, *Bull. WHO*, 38, 325, 1968.
13. **Falconer, D. S.,** *Introduction to Quantitative Genetics*, Oliver & Boyd Ltd., Edinburgh, 1964.
14. **Georghiou, G. P.,** Parasitological review, genetics of resistance to insecticides in houseflies and mosquitoes, *Exp. Parasitol.*, 26, 224, 1969.
15. **Finney, D. J.,** *Probit Analysis*, 2nd ed., Cambridge University Press, Cambridge, New York, 1952.
16. **Preisler, H. K., Hoy, M. A., and Robertson, J. L.,** Statistical analysis of modes of inheritance for pesticide resistance, *J. Econ. Entomol.*, 83, 1649, 1990.
17. **Halliday, W. R. and Georghiou, G. P.,** Inheritance of resistance to permethrin and DDT in the southern house mosquito (Diptera: Culicidae), *J. Econ. Entomol.*, 78, 762, 1985.
18. **Roush, R. T., Combs, R. L., Randolph, T. C., MacDonald, J., and Hawkins, J. A.,** Inheritance and effective dominance of pyrethroid resistance in the horn fly (Diptera: Muscidae), *J. Econ. Entomol.*, 79, 1178, 1986.
19. **Payne, G. T., Blenk, R. G., and Brown, T. M.,** Inheritance of permethrin resistance in the tobacco budworm (Lepidoptera: Noctuidae), *J. Econ. Entomol.*, 81, 65, 1988.
20. **Hoy, M. A. and Knop, N. F.,** Selection for and genetic analysis of permethrin resistance in *Metaseiulus occidentalis*: genetic improvement of a biological control agent, *Entomol. Exp. Appl.*, 30, 10, 1981.
21. **Strickler, K. and Croft, B. A.,** Selection for permethrin resistance in the predatory mite *Amblyseius fallacis*, *Entomol. Exp. Appl.*, 31, 339, 1982.
22. **Croft, B. A. and Whalon, M. E.,** Inheritance and persistence of permethrin resistance in the predatory mites *Amblyseius fallacis* (Acarina: Phytoseiidae), *Environ. Entomol.*, 12, 215, 1983.
23. **van de Klashorst, G.,** Acarine predator-prey interactions and their response to changes in components of the 'life system': a study of *Typhlodromus occidentalis* (Nesbitt) and *Tetranychus urticae* (Koch) on 'mini-orchards' in the glasshouse, Ph.D. thesis, Australian National University, 1983, 186.
24. **Petrushov, A. Z. and Zilbermints, I. V.,** Methods of obtaining and use of phytoseiid mites resistant to pesticides in USSR, in Programme VII Int. Congr. Acarology, Bangalore, India, August 1988.
25. **Petrushov, A. Z. and Gorshkova, E. V.,** Development of resistance of western *Metaseiulus* to Ambush, *Zashch. Rast.*, 9, 24, 1990.
26. **Hoy, M. A.,** Genetic improvement of phytoseids: in theory and practice, in *Modern Acarology*, Vol. 1, Dusbabek, F. and Bukva, V., Eds., SPB Academic Publishing, The Hague, 1991, 175.

Chapter 7

LONG-TERM CULTURING AND QUALITY ASSESSMENT OF PREDATORY BIG-EYED BUGS, *GEOCORIS PUNCTIPES**

Allen C. Cohen and Robert T. Staten

TABLE OF CONTENTS

* Mention of a tradename or specific product does not constitute an endorsement by the authors or the U.S. Department of Agriculture.

0-8493-2607-9/94/$0.00+$.50

I. INTRODUCTION

An important goal in augmentative biological control programs is establishing a system of rearing high-quality biocontrol agents that are economically feasible. It is essential that both the biological quality and the economic factors be suitable for a successful program. However, between these two factors, the economics may change under various circumstances such as reduced costs of materials as a result of large-scale purchasing of supplies, automation of handling, etc. By contrast, the biological quality of the control agents can never be compromised. However, concerns have been expressed by the National Academy of Sciences, Washington, D.C.,[1] about the potential for deterioration in the quality of entomophages reared on artificial hosts or prey. These concerns have been repeated frequently in the entomological community, so that there is a widespread distrust of products of artificial rearing. Such a decline in entomophage quality might be expected, especially of insects reared on artificial media. It is with this caveat in mind that we incorporated a regimen of quality assessment into our system of moderate-scale rearing of the predator *Geocoris punctipes* (Say) (Hemiptera: Lygaeidae).

The rearing of *G. punctipes* on an artificial diet[2] (AD) was begun in 1983, and the population which we describe here has been in continuous culture for nearly 10 years and nearly 100 continuous generations. Because of the unique character of this culture resulting from its long duration, it is very important to assess attributes of this population to answer questions of effects of long-term culturing upon entomophagous insects. The biology of *Geocoris* spp. has been studied by several authors.[3-6] Over the past 9.5 years, individuals from this culture have been subjected to a variety of assessments of their comparability to wild or insect-fed counterparts. Metabolic efficiency was measured.[7,8] Behavior tests were conducted to compare the laboratory-reared (LR) feeding responses and killing ability of *G. punctipes* with the same behavior parameters in wild counterparts.[8-10] Biological studies of the LR predators compared to wild counterparts are important in the assessment of the potential to use *in vitro*-produced entomophages. These factors include developmental rates, adult weights, and fecundity.[2,7,8,11]

Presented here are the results and discussions of a synthesis of these quality assessments. Several of these tests are compared with those reported earlier in this colony's history to determine possible changes that may have occurred over the nearly 100 generations of continuous artificial rearing. The major purpose of this work is to offer an objective assessment of several means of assessing the quality of laboratory-produced predators. Since this colony is, to our knowledge, the longest-standing culture of any entomophagous insect to be held on an artificial diet, we hope that this chapter will provide an objective basis for reconsidering the misgivings expressed by the National Academy of Sciences[1] and echoed frequently in the entomological community. The comparisons and observations made here should serve to further the understanding of predatory insects produced under other than natural conditions and provide a basis for assessment of entomophage quality. Such issues are fundamental to the entire concept of using *in vitro*-produced predators in any augmentative biological control of pests.

II. MATERIALS AND METHODS

A. COLONY MAINTENANCE AND CHANGES IN CULTURE TECHNIQUES USED IN EARLIER STUDIES

The original *G. punctipes* used to start the colony described here were collected in 1980 from an alfalfa field in the Avra Valley, 25 km northwest of Tucson, AZ. That original colony was fed coddled beet armyworms (*Spodoptera exigua* Hubner) and green beans (*Phaseolus vulgaris* L.) as described by Champlain and Sholdt.[6] A second colony was started from an alfalfa farm at Laveen, Maricopa County, AZ in 1991. A sample of the first colony was transferred to the artificial diet in December of 1983 and held on that diet since then, originally at the Arizona Biological Control of Insects Laboratory in Tucson and later at the Animal and Plant Health Inspection Service (APHIS), U.S. Department of Agriculture Methods Development Laboratory, Phoenix, AZ. The diet was the same composition as that reported by Cohen,[2] consisting of ground beef (21% fat), beef liver, and 10% sucrose solution in proportions of 45, 45, and 10%, respectively. The diet was blended with a food processor and dispensed from a hole in a plastic bag onto a Parafilm® sheet of about 200 × 8 cm. The sheet was folded so that the diet was completely covered with Parafilm. The Parafilm was sealed with a wooden rolling pin and cut into flat 8 × 8 cm feeding packets. After preparation, packets were kept frozen at –20°C until they were used. Thawed packets were stretched to about three times their original size and placed on the floor of the cage. Water was found to be a crucial factor in the survival of *G. punctipes* in culture, and until 1991, the colony was kept strictly on the artificial diet and free water provided with a sponge. However, since 1991, the colony at the APHIS facility has been kept both with water and green beans as extra insurance that water would be abundant and easily found by the predators, especially the first instar nymphs.

The cages now used to house cultures are formed of polyvinyl chloride plastic pressed into 30 × 21.5 × 6 cm (l × w × h) units glued onto cardboard. Lids are made of organdy-lined ice cream carton tops fitted onto a formed, raised circular top (13 cm diameter). The lid is easily removed and replaced to accommodate addition or removal of materials from cages. The organdy top allows ventilation of cages, but is fine enough to prevent escape of first instar nymphs.

Cultures were kept in a room held at 27 ± 1.5°C, 40% RH. The photoperiod was 14:10 light:dark phase. Eggs oviposited on a 10 × 10 cm cotton cloth are harvested daily and placed in new cages marked with the Julian day when they are collected. Each cage contained a population of *G. punctipes* of nearly identical age (eggs laid within 24 h of one another), thus creating age strata between but not within cages. The tightly timed age structure helped prevent large-scale disparities in the molting cycle. Such disparities are major factors in cannibalism observed in this species.[8]

B. BIOLOGICAL PERFORMANCE

The duration of each nymphal instar was determined by randomly selecting 30 individuals from each age stratum (each day) and counting the number of *G. punctipes* found to be in each instar. The weights of adults (3 to 5 days posteclosion) were determined for 3 groups of 20 adults each selected randomly from three cages. Individuals were lightly anaesthetized with CO_2 and weighed to the nearest 10 μg and sexed. Longevity was based on differences between Julian day of origin and current date. This was recorded four times during the course of 1992 and 1993. In August of 1986, approximately 3 years after cultures were started and after about 25 to 30 generations of continuous *in vitro* culturing, 44 adults (half of each sex) were compared with equal numbers of field-collected samples to determine changes in dry weight. Individuals were dried at 60°C and weighed to the nearest 0.1 mg. Means of dry weights were compared with a *t* test.

C. FIELD TESTS WITH Rb-MARKED PREDATORS

G. punctipes were marked with the rare element rubidium (Rb) so that captured individuals could be assessed for their origin as laboratory-released or wild predators.[12] Marking was achieved by incorporating Rb into the artificial diet at a concentration of 350 ppm. It was determined that such marking when started with last nymphal instars allowed detection of *G. punctipes* for 7 to 10 days after the predators were last fed on labeled diet. Prior to release, fields that were to be used as test sites were swept for *G. punctipes* that had originated in the field. Three groups of 50 adults were tested for Rb concentrations as described below for all capture/recapture subjects. Predators were released in one of four alfalfa fields in the Avra Valley at sites described above (each release comprising one replication). Each time, 460 newly eclosed adults were released into the center of the field. Each day after release, fields were sampled with 100 sweeps in each direction: north, south, east, and west or on alternate days northwest, southwest, northeast, and southeast. Sampling consisted of 100 sweeps in each direction, and the sweeps were continued for 100 m in each direction.

Collected insects were returned to the laboratory, aspirated, and frozen at –20°C within 1 h after capture. Specimens were dried at 65°C for 24 h, weighed to the nearest microgram and ashed in a muffle furnace at 750°C, taken up in a 1% solution of HNO_3, and analyzed according to the methods described by Cohen and Jackson[12] using a graphite furnace and a Perkin-Elmer 5000 atomic absorption spectrophotometer. Fecundity was measured in approximately the 90th generation AD-fed females over a 27-day posteclosion period. Twenty-five individuals of each sex were caged together 5 days after adult eclosion with food and water as described above and with cotton flannel patches for oviposition. Each day, eggs and females were counted, and tests were replicated three times.

In other field studies, LR *G. punctipes* (unmarked) were released into cages on either grapefruit trees in Imperial Valley, CA or on cotton plants in western Phoenix, AZ. The tests with grapefruit trees were conducted during spring of 1992, and releases of 5, 15, or 30 fourth and fifth instar predators were made into sleeve cages placed on the lower branches of trees on the south-facing side. Recaptures were done 1 week after the releases, and counts were made of the numbers captured and the number that had developed into the adult stage.

In the field tests with cotton, 460 adult or fifth instar *G. punctipes* were released into cages enclosing 1.5 row meters of severely infested cotton plants which were in their early fruiting stage. Prior to release, cages were carefully suctioned with a Devac apparatus to remove as many adult whiteflies and indigenous predators as possible. Also prior to release, three leaves were removed from each of five plants from each cage, leaf areas were measured with a LiCor area measuring apparatus, and 1.5-cm diameter punches were made near the base of each leaf. Eggs and nymphs of sweetpotato whiteflies were counted from each disk, and these counts were multiplied by the total surface area of each leaf to project the infestation of whiteflies per leaf. This number was multiplied by the mean number of leaves from each row foot of cotton in a comparable phenological stage.

III. RESULTS

A. CHANGES IN CULTURE TECHNIQUES

Several major changes were made to improve either the yield or the economy of rearing. (1) We returned to the use of green beans in all culture cages. Despite the fact that *G. punctipes* developed over 80 generations without plant matter,[8] we found that green beans attracted the predators, especially the early instars, to feed or drink for considerable time periods and in numbers often exceeding 5 to 10 individuals per bean. (2) We changed the diet packaging format from cylindrical "artificial larvae" to flattened packets. In both the old and new presentation techniques, stretched Parafilm was used to contain the diet. Various efforts to replace the Parafilm and to circumvent stretching were unsuccessful. (3) Diet packets were replaced daily rather than every other day as was done in earlier rearing.[2] This increase in frequency of feeding was started because of our observations of large aggregations of *G. punctipes* (>25 individuals per packet) on newly presented packets. These feeding aggregations lasted for about 2 to 4 h after new diet packets were presented, regardless of the time of day when they were offered. In the interest of stimulating feeding and, in turn, production and enhanced colony size, we made the adjustment to daily feeding. We felt that, especially in light of the fact that we use no preservatives or microbial inhibitors in the diet, a daily supply of fresh diet would be important. (4) The cotton wadding, used in earlier culture efforts as an oviposition site, was replaced with a cotton flannel material. This was a matter of material availability.

B. SURVEY OF THE LITERATURE

After comparing several parameters for quality assessment, it has become evident that judgments about the quality of AD-produced predators is highly variable and dependent upon which parameters are chosen to represent quality. For example, the metabolic, digestive, and absorption efficiencies of *G. punctipes* fed an artificial diet are considerably lower than the efficiencies of counterparts fed natural predator diets such as *Heliothis virescens*, *Helicoverpa zea*, or *Lygus hesperus* eggs.[7,8,11,13] Apparent digestibility of insect eggs was over 90%, and for artificial diet, it was slightly over 50%.[13,14] Efficiency of conversion of food to predator biomass was about 50 vs. 15% with insect eggs and artificial diet, respectively.[13,14] The higher metabolic efficiencies for predators consuming natural diets than those found with AD resulted from higher digestion and absorption efficiencies associated with natural prey rather than differ-

FIGURE 1. A *G. punctipes* from a culture that was reared approximately 90 generations on artificial diet. The predator is consuming *B. tabaci* on a cotton leaf.

ences in metabolic rates.[13-15] Cohen and Patana[16] showed that metabolic efficiencies are greater in insects feeding on natural diets than for those feeding on artificial diets, even including those insects that have been kept 25 years on an artificial diet.

G. punctipes from colonies reared on AD for over 6 years had feeding responses identical to those of wild *G. punctipes*.[9] In these tests, *L. hesperus* and *Acyrthosiphon pisum* (Harris) were preferred over *Aphis nerii* (Fonscolombe), *H. zea* (Boddie), *Heliothis virescens* (Fabricius), and *S. exigua* (Hubner). In paired comparisons of wild vs. AD-reared *G. punctipes* offered each type of prey, there were no differences in the acceptance rate by either predator type for any potential prey species. Figure 1 shows a 90th generation AD-fed *G. punctipes* feeding on *Bemisia tabaci* nymphs, demonstrating that there was no deterioration of predatory response.

IV. BIOLOGICAL CHARACTERISTICS OF THE PREDATORS

The development time for the various instars (Table 1) is about 3 days shorter in current production of the *G. punctipes* culture than it was in the culture's earlier phase (1 to 3 years after its origin). The weight of adult females was 5.4 (±0.51) in the current culture, and it was 5.4 (±0.39) in the culture nearly 10 years ago. The mean weights (±S.E.) of males was 3.4 (0.26) and 3.7 (0.35) in the current and earlier culture periods, respectively (Table 2). There were no significant differences in

TABLE 1
Development Time of *Geocoris punctipes* Fed Six Different Diets

	Duration of nymphal instar (days)					
Diet	First	Second	Third	Fourth	Fifth	Total
Helicoverpa zea eggs and beans[a]	5.1	3.3	4.3	4.1	5.8	22.7
Lygus hesperus eggs and beans[a]	4.8	3.9	4.5	4.3	6.1	23.6
L. hesperus eggs and H_2O[a]	5.1	4.1	4.3	4.3	6.4	24.2
H. zea eggs and H_2O[a]	5.4	4.8	4.1	4.9	7.6	26.8
Meat diet and H_2O[b]	6.1	6.2	5.0	4.2	7.1	28.6
Meat diet and beans and H_2O[c]	4.3	4.5	3.5	4.0	6.4	22.7

[a] Data from Cohen and Debolt.[11]
[b] Data from Cohen and Urias.[7]
[c] Data from present study.

TABLE 2
Weights of *Geocoris punctipes* from the Laboratory Compared with Weights from Field-Collected Samples (Means ± S.E.)

	Lab ♀♀	Field ♀♀	Lab ♂♂	Field ♂♂
Mean weights (mg)[a]	1.47 (0.042)	1.84 (0.044)	0.95 (0.023)	1.05 (0.034)

[a] Field females were significantly larger than AD-fed females (T = 5.14; $p < 0.05$; n = 22); field males were significantly larger than AD-fed males (T = 2.28; $p < 0.05$; n = 22).

weights of individuals from earlier and later cultures (Table 2). The sex ratios in the current culture was 50:50, as it was in the earlier culture period. The number of eggs per female in newly eclosed 90th generation AD-fed adults was about 2.5 and was not significantly different than the number produced by the 5th through 10th generation females.[11] Egg production by AD-fed females remains about half that of females produced on *H. virescens* eggs or *L. hesperus* eggs and water. Egg productivity slightly declined over a 27-day period (Figure 2).

A. Rb MARKING/FIELD RELEASES AND RECAPTURES

A total of 319 *G. punctipes* were captured in the four field releases, 47 of which were labeled with Rb, indicating that they originated from our cultures. Approximately 3% of the released predators were recaptured over a 10-day postrelease period. Since all recaptured insects were alive, it is a conservative estimate that at least 3% of all released insects survived, some of them surviving for at least 10 days in the field.

B. FIELD PERFORMANCE

In the experiments with *G. punctipes* being released into sleeve cages placed on citrus branches in Imperial Valley, CA during the spring of 1992, 92% of the total

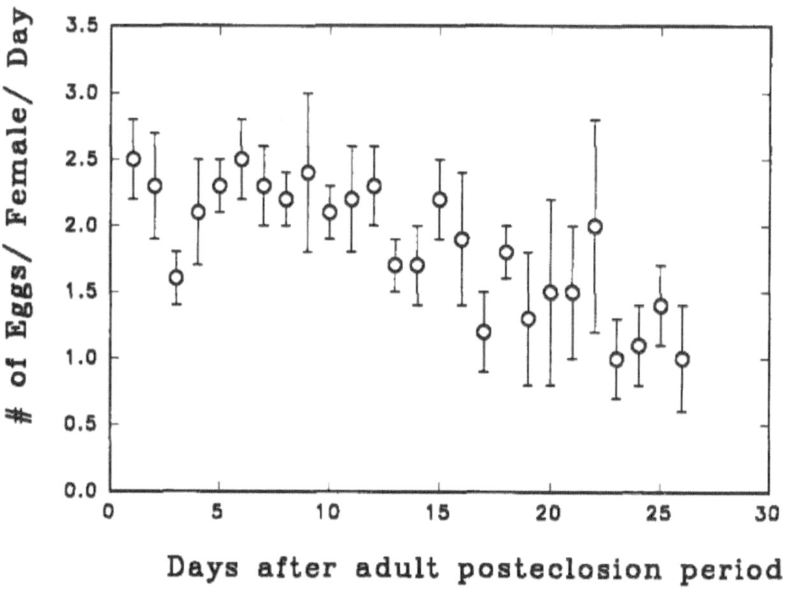

FIGURE 2. Daily oviposition rates of female *G. punctipes* from cultures reared nearly 100 continuous generations on artificial diet (means ± S.E.).

number released were recaptured 8 days after release. All individuals released were third instars; about 74% of the recaptures were fourth instars, and the rest were fifth instars. The only conspicuous food present within the cages was *B. tabaci*, mainly eggs and nymphal stages, and small amounts of California red scale. Evidently this material was nutritionally adequate to support the predators, allowing them to develop. It is also evident, and even more important, that the predators had not lost in over a decade of captivity the ability to survive in the field on natural prey.

In the second set of field cage experiments, after 10 days, we recovered nearly 52% of the predators released into the 3-m^3 cages over cotton plants in mid-August. In one of the five cages, we recovered over 80% of the number released, while our lowest recapture was about 28%. The population changes of nymphal and adult *B. tabaci* seen in Figure 3 show that there was a reduction in whitefly population in cages where predators were released. It is evident from this figure that whiteflies in cages with predator augmentation showed reductions in populations amounting to about one million eggs and one million nymphs per acre.

C. ECONOMIC EVALUATION

The actual economics of rearing these insects is speculative now because this is, in essence, still a laboratory colony. The current cost for our low- to moderate-scale rearing operation (producing less than 100,000 per week) are easily computed with minimum needs to estimate overhead, building expenses, etc. Costs are

1. Labor (feeding, cage preparation, diet preparation, etc.) at 42 h/week = $330.00
2. Supervision is $32.00/week

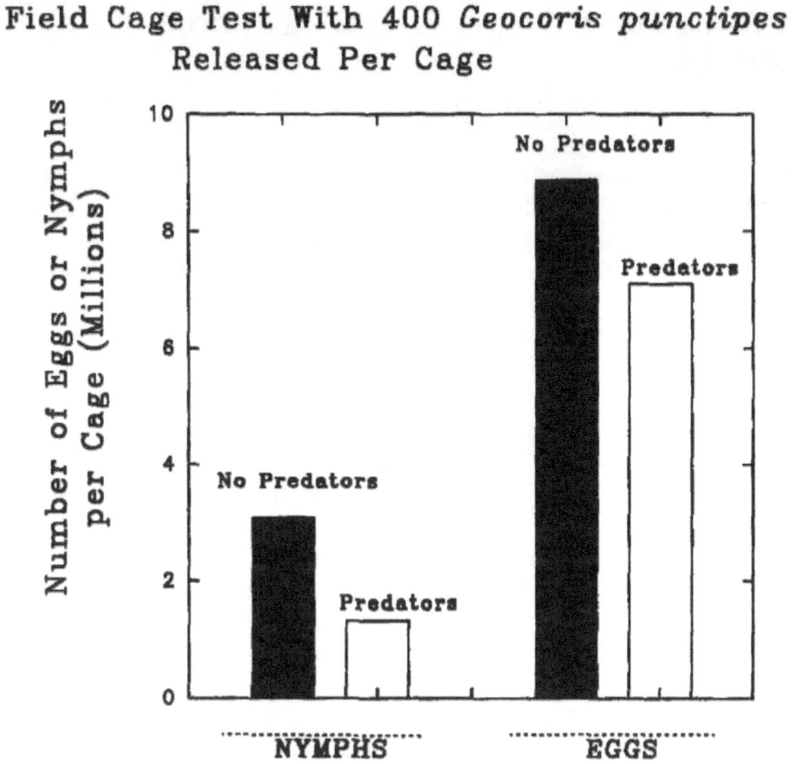

FIGURE 3. Effects of predation by *G. punctipes* on whitefly (*B. tabaci*) eggs and nymphs in field cages. The predators were reared for more than 90 generations on artificial diet.

3. Diet ingredients and all other rearing supplies are $76.00/week
4. Direct overhead at 20% is $87.00/week
5. Total cost per week is $525.00

The current production of third instar nymphs per week is 7000 to 8000. At the highest figure, these costs would be prohibitive for anything other than an inoculative release. Costs would be $65/1000. It is important to note, however, that 62% of all cost is in labor. Relatively simple automation of diet packets should cut this cost in half if colonies are to be run at a 100,000 level. At much higher levels, it is possible to reduce labor to 10% of the present level. Simply keeping all other figures constant, this cost would be $228/8000 per week representing a cost of $28/1000 insects. No profit is included. Informal price quotations on commercial first instar *Chrysoperla* spp. reared under large-scale production conditions (>5,000,000) have been $15 to 40/1000. Our experience (R. Staten, unpublished observations) in production of pink bollworm with relatively stable resource allocation and evolutional changes (improvement on an incremental basis) has led to near doubling of production from 1982 through 1991.[17] This, in effect, significantly lowers by 50% the cost of insects per 1000. It also illustrates the importance of understanding the dynamics of mass production of insects before simply looking at a cost per 1000 of relatively small-scale production.

V. DISCUSSION

King et al.[18] discussed the merits of using artificial diets to rear entomophages, including a greater potential for standardization and consistency of cultures, and reduction of rearing costs compared to the expense of rearing live hosts or prey. These potentially positive aspects of artificial rearing have been a driving force for several laboratories to undertake the painstaking efforts involved in developing artificial diets for entomophages. However, these efforts have always been undertaken with an undercurrent of suspicion that predators or parasites reared on other than natural hosts would be of inferior quality to their wild counterparts and would fail to be adequately competitive to carryout their expected roles as biological control agents. The core of this paradigm was stated in a position paper,[1] which appears to us to be based on little substance of an experimental nature. Certainly, at the time of writing that position paper, there were few instances where entomophages were being reared under artificial or unnatural circumstances; thus, there was very little basis for such a paradigm. Unfortunately, this paradigm has been repeated, and despite several evidences to the contrary discussed by King et al.,[18] there remains a specter of doubt about the value of unnaturally produced entomophages, especially those produced on artificial diet.

Over the 10 years of culture, there were some aspects that appeared to change and several others that appeared to remain constant; in some respects, the LR insects were indistinguishable from freshly collected individuals from the field. **Biological factors** showed the greatest differences between AD-fed predators and insect-fed or field-collected specimens. Adult weights were consistently lower in AD-fed specimens than they were in either field-collected or insect-fed predators. This factor neither improved nor degenerated after long-term culturing on AD. Development times were longer for AD-fed predators than they were for insect-fed counterparts early in the culture's history, but recent tests of later generations show a decrease in development time so that this factor is nearly identical for AD-fed and insect-fed predators. Longevity did not deteriorate over the period of AD feeding; individuals commonly lived for over 6x months in newer and older phases of the culture.

Metabolic and physiologic characteristics generally showed no signs of deterioration or improvement over the entire culture period, but the metabolic efficiency of food utilization and approximate digestibility measurements in AD-fed individuals was consistently lower than those measured in insect-fed predators. These measurements are probably the most direct means of assessing the dietary quality,[8] and they indicate that the materials that are of vertebrate origin (beef liver and ground beef) are not as digestible or nutritious as materials of insect origin, despite the fact the overall composition of proteins, carbohydrates, and lipids in the AD is nearly identical to the composition of an optimal diet derived from insect sources.[15] This fact, of course, confounds the purpose of entomophage artificial rearing — to provide foods that are less expensive and more convenient than diets composed of insect-derived materials. Metabolic rates, however, did not change in AD-fed insects.[8] It must be noted that the differences in metabolic efficiencies do not reflect permanent or genetic changes in the culture as is evidenced by the resumption of higher metabolic efficiencies by AD-cultured predators provided with insects as food.[8]

Behaviorally, individuals from long-standing cultures could not be distinguished from wild *G. punctipes*. Prey choices were identical for both groups of predators,[9] as were attack behaviors, feeding time, and use of stylet flanges.[8] **Ecologically**, it was evident that AD-reared predators could resume field activities, survive for at least 2 weeks, and reproduce in the field as evidenced by the field cage studies and Rb marking studies. This is a very important, fundamental point in light of concerns that predators that had become laboratory adapted might not be able to survive strenuous field conditions. It is especially impressive that some of the tests of field survival were conducted during the hottest part of the summer in Phoenix, AZ (in July and August). Even under this condition, the predators showed that they could survive and have some impact on prey populations (Figure 3).

The fact that this colony has been held under artificial rearing conditions for nearly 100 generations prompts some compelling questions about the quality of these predators, especially with respect to their capability in the field. The ultimate test of their ability to control pest population outbreaks would be an augmentative release of ample numbers of predators under circumstances such as a severe outbreak of a target pest or in experimental open field plots where controls without released predators show severe outbreaks. However, such a demonstration would require at least tens of millions of *in vitro*-produced predators, and production of these numbers would require substantial upscaling of the current rearing system. Such upscaling would be very costly in terms of labor, materials, and facilities; a major drawback of this expenditure has been doubts about the ability of predators reared under artificial conditions to perform satisfactorily. However, the extensive quality assessment efforts made over the past decade with *G. punctipes* as a subject should provide considerable assurance that at least this predator, reared under the conditions specified here, will be a useful agent of pest control in the field. We feel that there is strong efficacy shown here for the further development or upscaling of mass-rearing systems for *G. punctipes* based on 10 years of laboratory and field testing of behavioral, ecological, physiological, and metabolic aspects of LR specimens of this predator.

ACKNOWLEDGMENTS

We thank Drs. E. F. Knipling, F. M. Davis, S. M. Ferkovich, and A. A. Guerra for their helpful reviews of an earlier version of this manuscript and N. Colletto, G. Hafner, J. Eddy, L. Carland, S. Nardin, and P. Novak for their indispensable technical help.

REFERENCES

1. **National Academy of Sciences,** Control by parasites, predators, and competitors, *Insect-Pest Manag. Control*, 3, 100, 1969.
2. **Cohen, A. C.,** Simple method for rearing the insect predator *Geocoris punctipes* (Heteroptera: Lygaeidae) on a meat diet, *J. Econ. Entomol.*, 78, 1173, 1985.
3. **Crocker, R. L. and Whitcomb, W. H.,** Feeding niches of the big-eyed bugs *Geocoris bullatus, G. punctipes* and *G. uliginousus, Environ. Entomol.*, 9, 508, 1980.
4. **Dunbar, D. M. and Bacon, O. G.,** Feeding, development, and reproduction of *Geocoris punctipes* on eight diets, *Ann. Entomol. Soc. Am.*, 65, 892, 1972.
5. **Tamaki, G. and Weeks, R. E.,** Biology and ecology of two predators *Geocoris pallens* (Stal) and *G. bullatus* (Say), *U.S. Dep. Agric. Tech. Bull.*, 1446, 1, 1972.
6. **Champlain, R. A. and Sholdt, L. L.,** Rearing *Geocoris punctipes*, a lygus bug predator, in the laboratory, *J. Econ. Entomol.*, 59, 1301, 1966.
7. **Cohen, A. C. and Urias, N. M.,** Food utilization and egestion rates of the predator *Geocoris punctipes* fed artificial diets with rutin, *J. Entomol. Sci.*, 23, 174, 1988.
8. **Cohen, A. C.,** Using a systematic approach to develop artificial diets for predators, in *Advances in Insect Rearing for Research and Pest Management*, Anderson, T. E. and Leppla, N. C., Eds., Westview Press, New Delhi, 1985, 77.
9. **Hagler, J. R. and Cohen, A. C.,** Prey selection by *in vitro-* and field-reared *Geocoris punctipes, Entomol. Exp. Appl.*, 59, 201, 1991.
10. **Cohen, A. C. and Byrne, D. N.,** *Geocoris punctipes* as a predator of *Bemisia tabaci*: a laboratory evaluation, *Entomol. Exp. Appl.*, 65, 195, 1992.
11. **Cohen, A. C. and Debolt, J. W.,** Rearing *Geocoris punctipes* on insect eggs, *Southwest. Entomol.*, 8, 61, 1983
12. **Cohen, A. C. and Jackson, C. G.,** Using rubidium to mark a predator, *Geocoris punctipes* (Hemiptera: Lygaeidae), *J. Entomol. Sci.*, 24, 57, 1989.
13. **Cohen, A. C.,** Food consumption, food utilization and metabolic rates of *Geocoris punctipes* (Het.: Lygaeidae) fed *Heliothis virescens* (Lep.: Noctuidae) eggs, *Entomophaga*, 29, 361, 1984.
14. **Cohen, A. C.,** Ingestion efficiency and protein consumption by a heteropteran predator, *Ann. Entomol. Soc. Am.*, 82, 495, 1989.
15. **Cohen, A. C.,** Metabolic rates of two hemipteran members of a predator-prey complex, *Comp. Biochem. Physiol. A*, 81, 833, 1985.
16. **Cohen, A. C. and Patana, R.,** Efficiency of food utilization by *Heliothis zea* (Lepidoptera: Noctuidae) fed artificial diets or green beans, *Can. Entomol.*, 116, 139, 1984.
17. **Staten, R. T.,** Genetic control of cotton insects, the pink bollworm as a working program, *Proc. FAO/IAEA Int. Symp. Manag. Insect Pests: Nuclear Rel. Mol. Genet. Tech.*, 1993, 269.
18. **King, E. G., Hopper, K. R., and Powell, J. E.,** Analysis of systems for biological control of crop arthropod pests in the U.S. by augmentation of predators and parasites, in *Biological Control in Agricultural IPM Systems*, Hoy, M. A. and Herzog, D. C., Eds., Academic Press, Orlando, FL, 1985, 201.

Chapter 8

MAINTAINING GENETIC DIVERSITY IN LABORATORY COLONIES OF PARASITES AND PREDATORS

Alan C. Bartlett

TABLE OF CONTENTS

0-8493-2607-9/94/$0.00+$.50

Mother Nature may not always be right, but she insists on having her way.
Anonymous

I. INTRODUCTION

Although biological control concepts have been around much longer than 100 years (see Coppel and Merten[1] for an interesting history), for at least that long scientists have moved field-collected parasites, predators, and pathogens from one part of the world to another. The ultimate purpose of this movement is to introduce these natural enemies into the habitat of a pest species and to effectively eliminate (or, at least, suppress) the pest population in that habitat. Some of these introductions have been very successful, but many have not (for complete control, only 1 out of 17 attempts is successful).[2] Relatively few of these introductions have been accompanied by studies of the biology, genetics, and population structures of either (or both) of the contributing or the receiving populations. Unfortunately, we do not know if there is any clear correlation between the lack of such biological information and the success or failure of these introductions.

The introduction of natural enemies into a new area or against a new pest has been accomplished in two ways: **method 1** in which ample* numbers of the natural enemy are collected in the field (at some stage of development) and transported to the new area (after meeting quarantine regulations) where they are subsequently released (this does not happen much under new quarantine regulations); and **method 2** in which ample numbers of the natural enemy are mass reared in the laboratory for subsequent release. If a release is successful under either of these methods and the pest is eliminated, then we know that we released just the right number of the right natural enemies at just the right time.

If a release is not successful and the pest persists, then we conclude that insufficient numbers were released, it was not the right time for release, it was the wrong species to release, or the species we chose was not adapted to the release area. Often, if we conclude that the release numbers were insufficient to provide control using method 1, we again attempt to collect ample numbers of the natural enemy and take them into the laboratory to be mass reared. Thus, we revert to method 2. The maintenance of insect colonies has become an essential and valuable part of the biological control process. This chapter considers only the effects of mass rearing on the genetics of parasites and predators. I will not discuss the genetic effects of field collection and release, although some of the genetic consequences of such processes may be very similar to those which occur under a rearing procedure.

Mass rearing of natural enemies generally involves learning to rear two species, a host and a natural enemy (or even three species if a host plant is involved).**

* Ample is defined as the largest number of individuals we collect given the time and resources we have available.

** In a very few cases, natural enemies have been adapted to artificial diets. These cases have their own unique genetic consequences, but that is out of the scope of this chapter.

Additionally, in order for us to use natural enemies in an augmentative release, they must undergo both laboratory and field colonization. Therefore, I view the process of rearing in three dimensions. The first dimension is the genetic constitution of the insects; the second dimension is the native habitat; the third dimension is the laboratory habitat, including the host. It is an interplay of these three factors that determines the potential of a species to become established as a successful biocontrol agent.

II. GENETIC VARIABILITY

Modern genetic theory suggests that natural populations of organisms, including parasites and predators, have enormous amounts of genetic variability. Ayala,[3] for example, has suggested that at the level of the DNA sequence, every gene in every sexually reproducing organism is heterozygous. Although authors such as Owen[4] and Crespi[5] have found limited allozyme variability in certain hymenopterous parasites, recent studies of DNA variability[6] seem to validate Ayala's prediction. This has important implications for our study of genetic changes in laboratory-reared parasites and predators.

It has been estimated that higher organisms, e.g., insects, contain 4000 to 50,000 structural loci. We cannot study every one of these loci (even if the lowest estimate is correct), and we are probably not interested in the products of all of them. A diversity of genotypes may be successful under particular conditions. We should be interested in exploring the range of genetic variability in a population rather than examining only those polymorphisms that are easily detected.

A fundamental premise of the theory of evolution is that the presence of genetic variation in a population makes variation in survivability possible. Both natural and artificial selection rely on differential survivability to cause populational change. Adaptation is defined as the genetic process by which organisms undergo modification in nature to function more perfectly in a given environment. As a subset of adaptation, I define domestication (a process that requires human intervention)[7] as the genetic process by which organisms undergo modification to function more perfectly in the laboratory.

Recent evidence shows that there is ample genetic variation in natural populations to assure an opportunity for adaptation to occur,[8,9] even when samples of those populations are taken into the laboratory.[10,11] Even inbred populations can retain high amounts of genetic variation.[12] The problem is to determine how to maximize our ability to recognize, use, compensate for, or alter those adaptations which are involved in the characters critical to the success of biocontrol.

First, we must recognize how changes in gene frequencies can occur in a population. The following factors will cause changes in gene frequency over generations in the laboratory and will be considered here in this order.

1. The erosion of variation by sampling or selection
2. The input of new genetic material by mutation or immigration
3. The protection of variation by mating patterns or cytological mechanisms

A. THE EROSION OF VARIATION

When a population of parasites or predators is collected in the field, there is always a question about whether those traits that made this organism interesting or other traits that may affect performance will be lost due to sampling error. How many individuals should be obtained to minimize the chance of loss? The lower the number of insects taken into captivity, the lower the number of alleles that will be represented in the laboratory population.

Not every individual in a population will reproduce at the same rate. Some individuals will not reproduce at all, while others will reproduce at a high rate. Such differential reproduction reduces the effective number of parents that actually produce progeny. This has been clearly shown in many domesticated species. For example, in the 1930s, Holstein-Friesian cattle had more than 100,000 registrations per year, but the effective number of parents was estimated to be about 30.[13] Some of the difference between the effective number of parents and the number of individuals breeding is obviously due to artificial selection for the best bulls and cows. Similar heavy selection for productive parents may also take place naturally (although inadvertantly) during the laboratory domestication of insects.

Unruh et al.[14] showed that the effective number of parents in a laboratory population of *Aphidius ervi* may be as few as half the apparent number of individuals in that population. Studies with laboratory populations of other insects have found similar differences between the effective and actual number of parents.[15,16]

In an earlier paper on this subject,[17] I used the formula for the sampling variance of gene frequencies,

$$\sigma_g^2 = pq/2n_e$$

where σ^2 is the sampling variance of gene frequency, p is the frequency of one allele at a locus, q is the frequency of the second allele at a locus, and n_e is the effective number of parents. This shows that it provides more genetic diversity to start a population (assuming p = q = 0.5) with an effective number of 500 individuals (σ_g^2 = 0.0003) than with an effective number of 10 individuals (σ_g^2 = 0.0125). The larger the sampling variance, the greater the possible swing in gene frequency over generations or over sample populations. It is also very important to maintain population size over generations to keep gene frequencies relatively constant during the colonization process. Roush[18] concluded that collections of about 100 individuals from a given site (assuming that several sites are sampled) should preserve the native genetic heterozygosity in the laboratory population. We can estimate the sampling variance of gene frequencies for an effective number of 100 individuals to be σ_g^2 = 0.0013. Thus, the sampling variance of gene frequencies for a population is 9.6 times greater for 10 individuals than it is for 100 individuals and 4.3 times greater for 100 individuals than for 500 individuals. There is a diminishing advantage as numbers increase over an effective number of 100 individuals. Thus, the suggestion of Roush[18] is probably very close to the best number of individuals to use in a sample.

A similar conclusion can be reached by considering bottleneck effects on the average number of alleles per locus. Nei, Maruyama, and Chakraborty[19] used a

Monte Carlo simulation to see what happens to allelic frequencies that pass through bottlenecks. One of the original populations had 3.6 alleles per locus before the bottleneck. With bottleneck sizes of two individuals, there were averages of 1.3 alleles per locus remaining in the population. With a bottleneck size of 10 individuals, the average number of alleles per locus was 1.9, but the larger starting number yielded three times as many rare alleles as the lower one. Higher bottleneck sizes of the order of 50 to 100 individuals will effectively eliminate the bottleneck effect.[19]

Pimentel[2] suggested that 1000 or more vigorous individuals be collected to start a new colony to be tested as a biocontrol organism and that at least 1000 individuals be carried each generation as a stock population. However, Pimentel[2] gave no mathematical basis for the number he suggested. There is no doubt that the size of the collection should be as large as one can conveniently handle. However, the existence of island populations and the number of exotic pest species in our crops show that successful colonizations can occur with very low numbers of colonizers.

B. THE INPUT OF NEW VARIATION

According to classical genetic theory, new genetic variation can be introduced or returned to a colony by two processes: (1) mutation and (2) injection of new individuals into the colony by immigration. As we shall later see in this section, each of these processes involves some danger of disrupting the gene pool rather than simply introducing a new variation.

Spontaneous mutation in laboratory populations is usually ignored as a significant factor in the increase of genetic variability. However, I wonder if this assumption is appropriate. Most estimates of new mutations in higher organisms run on the order of 10^{-5} to 10^{-6} per locus per generation. The mutation rate of individual genes is low, but the number of new mutations in a population might be of some significance. For example, if we use an estimate of 27,000 loci per individual (this is the arbitrary arithmetic mean of the estimates of 4000 to 50,000 loci per individual) and assume diploidy, then, on the average, each individual might show $2 \times 27{,}000 \times 10^{-5} = 0.54$ mutations per generation or about 50 mutations per 100 individuals in the population. According to this calculation, the potential for changes in genetic variability due to spontaneous mutation could be a significant factor in laboratory populations.

Bartlett[20] suggested that new mutations could be introduced into laboratory populations of insects by mutagenic agents. Randomly induced mutation would be expected to increase genetic variance in the population. If selection for a particular trait is applied to the treated population, this increase in genetic variance should lead to a change in the mean of the character. This process has been used on a limited scale in plant breeding experiments.[21-23] These plant experiments have shown that polygenic and simply inherited mutations can be induced and utilized in breeding programs. In insects, experiments using mutagenic agents to induce genetic variability that can be used in selection procedures have also been successful.[24-26]

A disadvantage of the use of mutagenic agents to induce genetic variation is that recessive detrimental mutations, dominant lethal mutations, and chromosomal aberrations are likely to result from the treatment. If no selection is applied to the treated population to eliminate these detrimental characters, the average effect of the mutagenic treatment will lead to a reduction in fitness of all of the previously adapted genomes.[27] Such genetic changes will lower the reproductive fitness of the laboratory strain until those detrimental characters can be eliminated by rigorous selection for vigor and reproductive performance. Unfortunately, the selection process tends to reduce genetic variation, and so we are caught again in the cycle.

A suggested cure for the genetic differentiation of laboratory colonies from native populations is the regular introduction of native individuals into the colony. This process is considered in population genetics theory as "migration models" or "isolate breaking models".[28] Migration between populations sets a limit on the amount of genetic divergence which can occur between the populations. One effect of isolate breaking is the reduction in frequency of recessive homozygotes (assuming that the laboratory population has become inbred and that recessive homozygosis has increased) in the combined population.

It can be shown mathematically[7,29] that a change in gene frequency in the laboratory population will depend upon the ratio of the number of laboratory individuals to the number of introduced native individuals and on the difference in gene frequencies between the two populations. If there are no differences in gene frequency between the two populations, then it does not matter how many native insects are introduced. If the gene frequencies are different, then the direction of change will be toward the highest gene frequency, whether it is the laboratory frequency or the native frequency. Thus, we must know something about the genetic variabilities of the population before we can predict what sorts of changes might be expected to take place when native individuals are introduced into a laboratory population.

Several hazards can be associated with the introduction of native individuals into laboratory populations. For example, the introduction of pathogenic microorganisms poses a real threat to the health of a laboratory population. Also, outbreeding can lead to hybrid dysgenesis, depending upon the internal integration of the gene pools of the two populations.[30] Some studies have found that the strength of genetic incompatibility between two populations is proportional to the differences between environments where the races occur.[31,32] A similar correlation exists between the time that two populations have been separated and the amount of incompatibility between them.[33-35]

Although not commonly considered in classical population genetics theory, the techniques of molecular biology have opened new avenues for the introduction of genetic variation into laboratory populations. The introduction of bacterial genes into economically important plants has already been accomplished,[36] and production of transgenic livestock animals is not far behind.[37] Stable genetic transformation of a predatory mite, *Metaseiulus occidentalis*, has been accomplished.[38] Thus, it might be possible to either enhance the efficiency of biocontrol organisms or even to change their specificities by means of directed genetic change.[39,40]

It is possible that natural transposable elements could hasten the introduction of new genetic variation into laboratory populations of biocontrol organisms.[6,41] There is speculation that when organisms find themselves in unfamiliar niches, transposons may respond to the environmental stresses and increase mutation rates or increase the movement of genetic material from one species to another.[42]

C. THE PROTECTION OF VARIATION

There are forces in nature which can act to preserve variation in a population. For example, one of the fundamental assumptions in population genetics is that mating occurs randomly between prospective parents in a population. Under the assumption of random mating, no changes in gene frequency or genetic arrays occur, and the population is said to be in Hardy-Weinberg equilibrium. If the mating system changes, e.g., with inbreeding or outbreeding, gene frequencies are maintained, but genetic arrays will change. Homozygosity increases with inbreeding and decreases with outbreeding.

Breeding systems in populations can develop spontaneously or by the deliberate imposition by man. We expect that with low numbers of founders, inbreeding effects will follow. Steiner and Teig[43] found that in a selection program for increased insecticide resistance in *Microplitis croceipes*, many selected lines were lost because of an increase in the number of sterile diploid males. Inbreeding will increase homozygosity at the sex-determining locus in certain hymenopteran parasites and thus produce diploid males. In the same way, detrimental recessive genes may become homozygous and lead to a decline in the fitness or vigor of a population. Thus, individuals or even complete populations may be lost, even though allele frequencies may remain similar to the original population.

A large-scale experiment with *Drosophila pseudoobscura*, which was designed to investigate the founder-flush-crash model of speciation, found a definite bias toward positive assortative mating in populations which had few founders.[44] Such positive assortative mating could lead to rapid changes away from the original genetic composition of the population from which the founders came. Negative and positive assortative mating act very much like inbreeding and outbreeding in their effect on the genetic array of a population, except that the homozygosity (or heterozygosity) of only those loci affecting the assortative (disassortative) traits are changed.

Chromosomal aberrations can also have an effect on the retention of genetic variability. For example, paracentric chromosomal inversions (a change in the order of the genes on the chromosomes), chromosomal translocations (reciprocal exchanges of chromosome arms), and polyploidy all tend to keep blocks of genes together in the face of recombination. The heteromorphic sex chromosomes help to maintain polymorphisms in populations. The sex loci of many hymenopterous parasites are another case of biologically maintained polymorphisms.

III. NATIVE ENVIRONMENT

It is not possible to discuss every aspect of the native environment which could affect the behavior, adaptability, or availability of an arthropod under consideration for a biocontrol project. I am interested here in those environmental variables and their genetic interactions which could affect the outcome of biological control procedures that we might control. I am not sure that scientists can influence the availability of potential biocontrol organisms. In fact, investigators are often struck by the scarcity of potential parasites and predators of our most serious pest species.[45] Although it is not really in the domain of my discussion here, I believe that a phase of biocontrol experimentation should include the development of arthropod farms in areas of great biological diversity where both pest and biocontrol species could be cultured under natural conditions for use as reservoirs from which we can search for new candidate species. Native environments meeting these criteria are rapidly disappearing.

Roush,[18] in a comprehensive review of genetic variation in natural enemies, listed several attributes that might relate to success in a release program. For example, he suggested that attributes like searching capacity, dispersal, power of increase, prey consumption, and synchronization with host might be important to a biocontrol organism. However, he concluded that only two attributes (adaptability to local climatic conditions and the ability to avoid host defense) were commonly critical to success.

A recent review by Hoy,[46] concerning the contributions of Robert van den Bosch to biological control, emphasized that one of the major reasons for the successful introduction of 17 parasites into California was his emphasis on *appropriate biotypes* for each project. Roush[18] suggested that the operators of a biocontrol program could take steps to compensate for deficiencies in some attributes of a prospective biocontrol organism, but the ability of the organism to adapt to environmental changes and host defences is critical to biocontrol success. I think that Hoy (*vice* van den Bosch) and Roush are both emphasizing the same thing. Organisms will vary in their ability to exploit the resources available to them. We need to find the most successful exploiters relative to our particular environment and pest species.

One problem with the search for such successful organisms has been discussed by Pimentel.[2] Where hosts and parasites or predators have been in long association, they may have reached a point of equilibrium or balance of their respective populations. There is a balanced supply/demand economy in the system. At that point or in that system, it may appear that the pest species does not cause a serious problem and thus the biocontrol organism, which is balancing the equation, may not be useful on the pest in a new environment. The "new association hypothesis"[47] suggests that enemies of the target species from ranges other than the native range should be investigated. The most abundant natural enemies of a pest in the old environment are often the ones chosen for use in the new area. Hokkanen[47] suggests that rare beneficial enemies of the target species in the native environment may be superior in their ability to exploit the new environment because of the new association phenomenon.

It seems obvious that the choices of potential biocontrol agents should be accompanied by detailed biological studies of the organism in its native environment. However, such studies are not commonly done. Funds are available for collection and importation of arthropods for biocontrol purposes, but studies of the same arthropods in their native environment are scarce.[48] Further, detailed studies of biocontrol organisms under laboratory conditions sometimes lead to erroneous conclusions.[49]

We know that arthropod characteristics can vary over space and time. The abilities of biocontrol organisms to adjust to the spatial and temporal variation of both host and environmental conditions are particularly cumbersome for detailed study. Studies of clinal variation in parasites and predators should be of great interest in the selection of species appropriate to a given host environment, but evidence of such

TABLE 1
Important Environmental Conditions Which May Differ Between Field and Laboratory Populations of Biocontrol Organisms

Field	Laboratory
Hosts scarce	Hosts abundant
Hosts dispersed	Hosts proximate
Mates scarce	Mates abundant
Mates dispersed	Mates proximate
Temperature fluctuates	Temperature optimal
Humidity fluctuates	Humidity optimal
Light fluctuates	Light optimal
Densities fluctuate	Density crowded

studies is difficult to find (see Chapter 10 for particular examples of the appropriate selection of biocontrol organisms). However, such information is most pertinent in situations when these biocontrol organisms are called upon to attack a new host in a new environment.[2]

IV. LABORATORY ENVIRONMENT

Proper handing of the first generation in rearing is crucial because the existing gene frequency depends not only on the number of individuals that start the colony, but also on the previous native environmental history. In other words, certain genes are frequent in the native population because of the effects of natural selection and genetic drift in the native environment. However, survival of the new colony depends upon the present laboratory environment and the way that the sample population gene pool responds to that new environment. The colony will be fit in both environments only if a strong positive correlation exists between the two environments. The challenge for the biocontrol worker is to optimize rearing conditions to maintain that strong positive correlation.

We can conceive that genetic changes could occur due to natural selection when an organism is placed into a laboratory environment, even if no conscious selection is taking place. Table 1 lists some differences between the native environment and a common laboratory environment which could cause selective changes to occur. An additional problem, in the case of parasites and predators, is that even the host organism may have been previously laboratory adapted or may not be the normal host of the beneficial organism. Biological control workers do not seem to be overly concerned with changes in host preferences due to rearing on unnatural hosts,[50] but perhaps they should be.

In the laboratory, environmental parameters are usually made optimum for the worst genotype in the population. We want individuals that will lay eggs in a place convenient for our collection methods. We do not want the captive organisms to fly or run away when we are trying to examine them or when we are removing the host or food sources. In the field, a female may pass up several host larvae to find what she considers a suitable host. We want her to use every host larvae we provide, and we favor those females that are less discriminating. In short, we want to produce maximum numbers with minimum input.[51]

Thus, by unconscious selection, we further erode the genetic variation which survived our sampling efforts. Selection is also much more efficient in effecting changes in gene frequencies than chance phenomena because it is usually directional. A large effective number of parents affects response to selection by making the selection process even more efficient. To reduce the effects of unconscious selection, we must reduce the differences between the native and the laboratory environment. We must try to preserve those characters which appear to be important in the establishment of predators and parasites in new native environments. Several investigators have suggested the following to be important (see the list in Caltagirone[52] for other examples):

1. Ecological specialization, including a narrow range of hosts and synchronous with host biology
2. High powers of dispersal, discovery, and persistence
3. High rate of fecundity at times of host expansion
4. Rapid developmental rate
5. Capability to coevolve with host, e.g., capability to acquire pesticide resistance or to adapt to shifts in the host's defense mechanisms

I think that biocontrol workers shall never see a biocontrol organism that meets all these criteria. There is another problem. It is not in the nature of a normal biological enemy to eliminate its host. In fact, to do so would be ecological suicide, but we often intend for our augmentative release programs to accomplish this. We may really need to design "natural enemies" according to our specifications rather than to those acceptable to natural selection (may I say — Mother Nature?). I believe that one of the best contributions of Hoy[46,48] to the field of biocontrol has been her insistence and demonstration that genetic techniques can and do work and should be used to improve the establishment of natural enemies. As more examples of genetic research appear in the literature, the improvement of rearing procedures will also follow. It is not practical to make genetic crosses without a successful method of rearing the progeny of those crosses in the laboratory.

V. CONCLUSIONS

In evolutionary history, low numbers of colonists have successfully invaded new territories. In biological control history, low numbers of introduced natural enemies have successfully controlled pest species. In both cases, there had to be a favorable niche to which the organism was either preadapted or could rapidly become adapted. The genetic plasticity of biocontrol organisms seems almost without limit. Surely, using our knowledge of population genetics, molecular genetics, entomology, and natural history, we can study and utilize that plasticity to produce laboratory strains of natural enemies suited to the various needs of agriculture.

Biocontrol workers should follow a few simple rules in domesticating biocontrol agents.

1. Obtain as much biological information about the native environment of the host as possible, and see what array of natural enemies exist.
2. Collect not only those natural enemies most prevalent in the native environment, but also those that are less successful or at the outer range of the host distribution.
3. Collect as least 100 specimens of each natural enemy for establishment of the colony, and try to obtain specimens from various niches that exist within the natural range of the host.
4. In the first few generations of rearing, keep careful records of the numbers of reproducing adults to maintain a broad base of effective numbers of parents. In other words, do not allow only a few females to provide all of the progeny from which a colony is established. Try to avoid genetic drift or any peculiar breeding system such as assortative mating.
5. Try to make the laboratory environment as variable as practical. Make the natural enemy work a little to find the host. Vary the light, temperature, humidity, etc. in a natural diel pattern. Try to avoid natural selection for the equivalent of a laboratory couch potato.
6. If the natural enemy normally has some unique behavioral characteristic (for example, goes into diapause in the host), try to maintain that character in the colony.
7. If inbreeding depression or extremely skewed sex ratios occur, attempt to add new genetic variation to the population by introducing new native insects.
8. Understand that this domesticated strain does not have to have all of the characteristics of native individuals, because they may be used in augmentative releases and do not necessarily have to become established and survive forever in the field.[53]

REFERENCES

1. **Coppel, H. C. and Mertins, J. W.,** Historical development, in *Biological Insect Pest Suppression*, Springer-Verlag, New York, 1977, chap. 2.1.
2. **Pimentel, D.,** Population dynamics and the importance of evolution in successful biological control, in *Handbook of Pest Management*, Vol. 2, Pimentel, D., Ed., CRC Press, Boca Raton, FL, 1990, 171.
3. **Ayala, F.,** Evolutionary significance of genetic variation in insects, in *The Evolutionary Significance of Insect Polymorphisms*, Stock, M. W. and Bartlett, A. C., Eds., University of Idaho Press, Moscow, ID, 1982, 1.
4. **Owen, R. E.,** Difficulties with the interpretation of patterns of genetic variation in the eusocial Hymenoptera, *Evolution*, 39, 201, 1985.
5. **Crespi, B. J.,** Heterozygosity in the haplodiploid Thysanoptera, *Evolution*, 45, 458, 1991.

6. **Cockburn, A. F., Howells, A. J., and Whitten, M. J.,** Recombinant DNA technology and genetic control of pest insects, in *Genetical and Biochemical Aspects of Invertebrate Crop Pests*, Russell, G. E., Ed., Intercept, Andover, U.K., 1989, 319.
7. **Bartlett, A. C.,** Genetic process of domestication, in *Advances and Challenges in Insect Rearing*, King, E. G. and Leppla, N. C., Eds., U.S. Department of Agriculture Handbook, U.S. Department of Agriculture, New Orleans, LA, 1984, 2.
8. **Hung, A. C. F., Hedlund, R. C., and Day, W. H.,** High level of genetic heterozygosity in the hyperparasitic wasp, *Mesochorus nigripes*, *Experientia*, 42, 1050, 1986.
9. **Hung, A. C. F., Day, W. H., and Hedlund, R. C.,** Genetic variability in arrhenotokous and thelytokous forms of *Mesochorus nigripes* (Hym.: Ichneumonidae), *Entomophaga*, 33, 7, 1988.
10. **Bartlett, A. C.,** Isozyme polymorphisms in populations of the pink bollworm, *Ann. Entomol. Soc. Am.*, 74, 9, 1981.
11. **Tsakas, S. C. and Zouros, E.,** Genetic differences among natural and laboratory-reared populations of the olive-fruit fly *Dacus oleae* (Diptera: Tephritidae), *Entomol. Exp. Appl.*, 28, 268, 1980.
12. **Yamazaki, T.,** Detection of single gene effect by inbreeding, *Nature (London)*, 240, 53, 1972.
13. **Lush, J. L.,** *Animal Breeding Plans*, Iowa State College Press, Ames, 1937.
14. **Unruh, T. R., White, W., Gonzalez, D., Gordh, G., and Luck, R. F.,** Heterozygosity and effective size in laboratory populations of *Aphidius ervi* (Hym. Aphidiidae), *Entomophaga*, 28, 245, 1983.
15. **Kerr, W. E. and Wright, S.,** Experimental studies of the distribution of gene frequencies in very small populations of *Drosophila melanogaster*, *Evolution*, 8, 172, 1954.
16. **Prout, T.,** Genetic drift in irradiated experimental populations of *Drosophila melanogaster*, *Genetics*, 39, 529, 1954.
17. **Bartlett, A. C.,** Guidelines for genetic diversity in laboratory colony establishment and maintenance, in *Handbook of Insect Rearing*, Singh, P. and Moore, R. F., Eds., Elsevier Science Publishers B.V., Amsterdam, 1985, 7.
18. **Roush, R. T.,** Genetic variation in natural enemies: critical issues for colonization in biological control, in *Critical Issues in Biological Control*, Mackauer, M., Ehler, L. E., and Roland, J., Eds., Intercept, Andover, U.K., 1990, 263.
19. **Nei, M., Maruyama, T., and Chakraborty, R.,** The bottleneck effect and genetic variability in populations, *Evolution*, 29, 1, 1975.
20. **Bartlett, A. C.,** Genetic markers, in *Sterile Insect Technique and Radiation in Insect Control*, IAEA-SM-255, International Atomic Energy Agency, Vienna, 1982, 451.
21. **Mackill, D. J. and Rutger, J. N.,** The inheritance of induced-mutant semi-dwarfing genes in rice, *J. Hered.*, 70, 335, 1979.
22. **Krull, C. F. and Frey, K. J.,** Genetic variability in oats following hybridization and irradiation, *Crop Sci.*, 1, 141, 1961.
23. **Brock, R. D. and Latter, B. D. H.,** Radiation-induced quantitative variation in subterranean clover, in *Radiobiology*, Butterworths, London, 1961, 205.
24. **Scossiroli R. E. and Scossiroli, S.,** On the relative role of mutation and recombination in responses to selection for polygenic traits in irradiated populations of *D. melanogaster*, *Int. J. Radiat. Biol.*, 1, 61, 1959.
25. **Bartlett, A. C., Bell, A. E., and Anderson, V. L.,** Changes in quantitative traits of *Tribolium* under irradiation and selection, *Genetics*, 54, 699, 1966.
26. **Aastveit, K.,** Induced mutations in polygenic systems, *Hereditas*, 56, 20, 1966
27. **Wallace, B.,** *Genetic Load*, Prentice-Hall, Englewood Cliffs, NJ, 1970.
28. **Hartl, D. L. and Clark, A. G.,** Population subdivision and migration, in *Principles of Population Genetics*, 2nd ed., Sinauer Associates, Inc., Sunderland, MA, 1989, chap. 6.
29. **Pirchner, F.,** *Population Genetics and Animal Breeding*, W. H. Freeman, San Francisco, 1969.
30. **Spiess, E. B.,** Integration and separation of gene pools, in *Genes in Populations*, John Wiley & Sons, New York, 1977, chap. 19.
31. **Jaenson, T. G. T.,** Mating behavior of *Glossina pallidipes* Austen (Diptera, Glossinidae): genetic differences in copulation time between allopatric populations, *Entomol. Exp. Appl.*, 24, 100, 1978.

32. **Jansson, A.,** Viability of progeny in experimental crosses between geographically isolated populations of *Arctocorisa carinata* (C. Sahlberg) (Heteroptera: Corixidae), *Ann. Zool. Fenn.*, 15, 77, 1978.
33. **Richerson, V. and Cameron, E. A.,** Differences in pheromone release and sexual behavior between laboratory-reared and wild gypsy moth adults, *Environ. Entomol.*, 3, 475, 1974.
34. **Raulston, J. R., Graham, H. M., Lingren, P. D., and Snow, J. W.,** Mating interaction of native and laboratory-reared tobacco budworms released in the field, *Environ. Entomol.*, 5, 195, 1976.
35. **Sharp, J. L.,** Comparison of flight ability of wild-type and laboratory-reared fruit flies on a flight mill, *J. Ga. Entomol. Soc.*, 11, 255, 1976.
36. **Gasser, C. S. and Fraley, R. T.,** Genetically engineering plants for crop improvement, *Science*, 244, 1293, 1989.
37. **Pursel, V. G., Pinkert, C. A., Miller, K. F., Bolt, D. J., Campbell, R. G., Pamiter, R. D., Brinster, R. L., and Hammer, R. E.,** Genetic engineering of livestock, *Science*, 244, 1281, 1989.
38. **Presnail, J. K. and Hoy, M. A.,** Stable genetic transformation of a beneficial arthropod, *Metaseiulus occidentalis* (Acari: Phytoseiidae), by a microinjection technique, *Proc. Natl. Acad. Sci. U.S.A.*, 89, 7732, 1992.
39. **Beckendorf, S. K. and Hoy, M. A.,** Genetic improvement of arthropod natural enemies through selection, hybridization or genetic engineering techniques, in *Biological Control in Agricultural IPM Systems*, Hoy, M. A. and Herzog, D. C., Eds., University of California Press, Berkeley, 1985, 167.
40. **Whitten, M. J.,** The relevance of molecular biology to pure and applied entomology, *Entomol. Exp. Appl.*, 53, 1, 1989.
41. **Handler, A. M. and O'Brochta, D. A.,** Prospects for gene transformation in insects, *Annu. Rev. Entomol.*, 36, 159, 1991.
42. **Amabile-Cuevas, C. F. and Chicurel, M. E.,** Horizontal gene transfer, *Am. Sci.*, 81, 332, 1993.
43. **Steiner, W. W. M. and Teig, D. A.,** *Microplitis croceipes* (Cresson): genetic characterization and developing insecticide resistance biotypes, *Southwest. Entomol.*, 14(Suppl. 12), 71, 1989.
44. **Galiana, A., Moya, A., and Ayala, F. J.,** Founder-flush speciation in *Drosophila pseudoobscura*: a large-scale experiment, *Evolution*, 47, 432, 1993.
45. **Pimentel, D.,** Population regulation and genetic feedback, in *Natural Regulation of Animal Populations*, McLaren, I. A., Ed., Atherton Press, New York, 1971, 171.
46. **Hoy, M. A.,** Biological control in U.S. Agriculture: back to the future, *Am. Entomol.*, 39, 140, 1993
47. **Hokkanen, H. M. T.,** New approaches in biological control, in *Handbook of Pest Management*, Vol. 2, Pimentel, D., Ed., CRC Press, Boca Raton, FL, 1990, 185.
48. **Hoy, M. A.,** Improving establishment of arthropod natural enemies, in *Biological Control in Agricultural IPM Systems*, Hoy, M. A. and Herzog, D. C., Eds., Academic Press, New York, 1985, 151.
49. **Tumlinson, J. H., Lewis, W. J., and Vet, L. E. M.,** How parasitic wasps find their hosts, *Sci. Am.*, 266 (3), 100, 1993.
50. **King, E. G. and Morrison, R. K.,** Some systems for production of eight entomophagous arthropods, in *Advances and Challenges in Insect Rearing*, *King*, E. G. and Leppla, N. C., Eds., U.S. Department of Agriculture Handbook, U.S. Department of Agriculture, New Orleans, LA, 1984, 206.
51. **King, E. G., Hopper, K. R., and Powell, J. E.,** Analysis of systems for biological control of crop arthropod pests in the U.S. by augmentation of predators and parasites, in *Biological Control in Agricultural IPM Systems*, Hoy, M. A. and Herzog, D. C., Eds., Academic Press, New York, 1985, 201.
52. **Caltagirone, L. E.,** Identifying and discriminating among biotypes of parasites and predators, in *Biological Control in Agricultural IPM Systems*, Hoy, M. A. and Herzog, D. C., Eds., Academic Press, New York, 1985, 189.
53. **Beckendorf, S. K. and Hoy, M. A.,** Genetic improvement of arthropod natural enemies through selection, hybridization or genetic engineering techniques, in *Biological Control in Agricultural IPM Systems*, Hoy, M. A. and Herzog, D. C., Eds., Academic Press, New York, 1985, 167.

Chapter 9

HOST-RANGE INCREASE OF INDIGENOUS AND INTRODUCED PARASITOIDS

Gennaro Viggiani and Dan Gerling

TABLE OF CONTENTS

0-8493-2607-9/94/$0.00+$.50

I. INTRODUCTION

A parasitoid, in order to oviposit, must undergo a series of steps, starting with host habitat location, continuing through host location within the habitat, and culminating in host recognition and acceptance.[1-3] Furthermore, in order for its progeny to develop successfully, host manipulation by the female as well as physiological compatibility on the part of the developing parasitoid with the host's biochemistry, physiology, and defense mechanisms are all essential.[4,5]

Thus, a specialist parasitoid may cue in on very specific host-associated features or characteristics, and will have very close physiological and biochemical association with its host. Such features will facilitate a more efficient host exploitation than that of other parasitoid species, enabling it to outcompete them. However, the danger in such specialization rests in the extreme dependence that the specialized parasitoid develops toward a specific host (or even a host stage or a certain host associated habitat) (e.g., see Reference 6).

An oligophagous parasitoid, like some *Encarsia* species, still shows a considerable degree of specialization without having limited itself to a single species and/or stage. This is possible because host-searching mechanisms may involve an element of learning by the parasitoid,[7] while the adaptations necessary for physiological host/parasitoid compatibility may be general enough to apply to a number of species.

Like all natural mechanisms, host/parasitoid relations are dynamic processes. Consequently, we must assume that the present host range of any oligophagous species has not necessarily been the same in the past and will not remain stationary in the future. This change in host range involves the losing of some hosts (e.g., through extinction or the build-up of immunity) and the adoption of new ones.

The host ranges of many parasitoid species have been artificially enlarged by laboratory rearing in which the host-finding stages were circumvented.[8] The new or "factitious" laboratory hosts that were used in these cases must have been physiologically suitable for normal parasitoid development and usually belonged to the same, or close, taxonomic entity.

The process of parasitoids moving to new hosts in nature depends upon their flexibility (the degree to which they adapt to changes in host characteristics in a favorable way) and upon the availability of appropriate hosts. Modern transport systems have facilitated the movement of both hosts and parasitoids from one country to another. While the introduction of pests is often inadvertent, parasitoids are usually deliberately introduced in attempts to suppress the introduced pests in their new habitats. Under these conditions, new host ranges are noted during the process of parasitoid introduction and study. Moreover, the exposure of the parasitoid to a gamut of new potential hosts that were formerly allopatric in their ranges offers a unique possibility to observe host-range changes.

The literature has many examples of parasitoid host ranges. Their ability to adopt new hosts has been discussed and used to develop a theory about the superior potential of such new associations for biological insect control.[9,10] However, a full analysis of host-range changes necessitates exact taxonomic and ecologic identification of all three trophic levels: host plant, host insect, and parasitoid. Unfortunately, the current taxonomic knowledge concerning many of the groups involved, and in particular the parasitoids, is still incomplete, thus preventing determination of the entire range of host/parasitoid relationships.

During the last two decades, we have studied the parasitic fauna associated with the homopterous family Aleyrodidae (whiteflies). Several species of these plant pests were inadvertently transferred from one country to another, and, following them, their parasitoids were often deliberately introduced into new regions. Drawing from our experience, we shall discuss host-range changes, in particular those associated with the genera *Cales, Encarsia*, and *Eretmocerus*. The topics of our discussion will include adaptations by indigenous parasitoids to newly introduced hosts, by exotic parasitoids to the hosts available in a new area, and to new hosts by heteronomous parasitoids. Finally, we shall indicate cases of host resistance to range increases by the parasitoids.

II. ADAPTATION BY INDIGENOUS PARASITOIDS TO ACCIDENTALLY INTRODUCED HOSTS

Host-range changes occur constantly, many of them prompted by the new availability of a host. For example, the housefly *Musca domestica* L. was unknown in the Western hemisphere in pre-Columbian times, but it currently supports a considerable fauna of parasitoids there belonging to the genera *Spalangia* and *Muscidifurax*.[11] The parasitoids belonging to the latter genus originated in the Western hemisphere, adapted to the new hosts, and have probably undergone speciation while exploiting their new host association. They formed a species complex that is presently specific to the eusynanthropic fly habitat.[11]

A more recent and well-recorded example involves *Macrocentrus ancylivorous* Rohwer. This braconid is an endemic species in the eastern U.S., attacking the strawberry leaf roller *Ancylis compactana* (W & R). When the oriental fruit moth *Cydia molesta* (Busck) became established in the U.S. (*circa* 1913), it was readily attacked by the native *M. ancylivorous,* which was found to be well adapted to it in many areas. This became one of the most impressive examples of a native parasitoid proving effective against an introduced pest attacking a previously unaffected crop.[1]

In another example, the soft-scale parasitoid *Metaphycus helvolus* (Compere) was introduced from Africa into California to combat the scale *Saissetia oleae* (Olivier). However, shortly following its introduction, it also provided good control of the Nigra scale *S. nigra* (Nietner).[12]

Attempts to achieve biological control of whiteflies provided several examples of continual increases in the range of parasitoids. Of the nine species of parasitoids examined, five adapted to new hosts that had been introduced recently. For example, *Encarsia tricolor* Foerster, *E. inaron* (Walker) (= *partenopea* Masi), and *E. lutea* (Masi), all native Italian species, have adopted the greenhouse whitefly *Trialeurodes vaporariorum* (Westwood). The greenhouse whitefly is a native of the U.S.,[13] but was accidentally introduced into Europe, where it became a severe pest. In some places, mainly outdoors, *E. tricolor* has become an efficient controlling agent.[14]

E. luteola Howard (= *deserti* Rivney and Gerling), *E. pergandiella* Howard, and *E. formosa* Gahan are natives to the American continent. In 1966, the former was found attacking the sweetpotato whitefly *Bemisia tabaci* (Gennadius), an introduced pest from the Orient.[15] The latter two species have recently been reported parasitizing *B. tabaci*, following widespread outbreaks in the U.S. *E. tricolor* was also recently found to attack *B. tabaci*, at least in experimental set-ups.[15a]

The bayberry whitefly *Parabemisea myricae* (Kuwana) of Asian origin invaded southern California in 1978. Several parasitoids were imported into California from Japan, mass reared, colonized, and established,[16] but *Eretmocerus debachi* (Rose and Rosen), a species that had not been deliberately introduced and is presumed to be native to California, proved to be the most effective control in southern California (and later also in several Mediterranean countries).[16,17]

III. ADAPTATION OF EXOTIC PARASITOIDS TO THE HOSTS AVAILABLE IN A NEW AREA

As noted, the importation of parasitoids into new areas or countries exposes them to new hosts. While these often belong to the same or very close taxa as their original hosts, there is clearly a change in their host range.

Encarsia transvena (Timberlake) has invaded numerous areas, including Hawaii and Kenya, where it has become a prominent parasitoid of *T. vaporariorum* and *B. tabaci*.[18,19] This situation is interesting because *T. vaporariorum* was not known as a host of *E. transvena* in Asia, where this parasitoid apparently originated. *B. tabaci*, one of the probable Asian hosts of *E. transvena*, did not occur at the time in Hawaii. *E. transvena* probably arrived in Hawaii as a parasitoid of *Singhius hibisci* (Kotinsky) and, once there, adapted to *T. vaporariorum*. The latter invaded Hawaii from the New World in which *E. transvena* was unknown at the time.[18] *E. transvena* has since spread widely and occurs as a parasitoid of both *T. vaporariorum* and *B. tabaci* in several African countries.[19] It has also been introduced into Israel,[19a] as well as into Italy, where it has been recorded from two new whitefly hosts, *Asterobemisia carpini* (Koch) (= *avellanae* (Signoret) and *P. myricae* (Kuwana).[20]

A Nearctic species, *E. pergandiella*, introduced into Italy to improve the biological control of *T. vaporariorum*, spread quickly from the areas of release and became established, moving from the north to the south of the country.[21] It also adapted to four other whitefly species, a fact that helps to explain its spectacular spread and successful colonization.[22,23]

E. meritoria Gahan, which is also of Nearctic origin, was first recorded in the Mediterranean basin from Spain and Italy, where it parasitized the greenhouse whitefly.[24,25] Since then, it has been recorded from two additional hosts in Italy, *Aleyrodes lonicerae* Walker and *P. myricae* (Kuwana), none of which originated in the Western hemisphere.

The woolly whitefly *Aleurothrixus floccosus* (Maskell), with an original distribution that included the West Indies and South and Central America,[26] became established in southern California in 1966. The aphelinid *Cales noacki* Howard, whose known host range at the time included only species of the genus *Aleurothrixus*, was introduced into California and later into Europe in a very successful campaign for the biological control of the woolly whitefly.[27]

Since then, *C. noacki* has adopted numerous new hosts, among the most important of which is the red-banded whitefly *Tetraleurodes* sp. in California. When *Tetraleurodes* sp. infested avocado in San Diego county in September 1982, a survey was carried out for its natural enemies. The results of the initial searches were negative, but since November 1982, *C. noacki* was found to parasitize the whitefly successfully. Rates of parasitism increased during the winter of 1983 to an overall mortality rate of about 92%, which resulted in substantial biological control of the pest.[28]

After its introduction into Europe, *C. noacki* was recorded on several hosts. Shortly after its introduction into Italy, it was obtained once from eggs of *Phalera bucephala* L. (Lepidoptera: Notodontidae)[29] and later from *Aleurotuba jelinekii* (Frauenfeld), a univoltine whitefly,[30] where it gradually displaced *E. margaritiventris* (Mercet),[31] reaching a 80 to 90% rate of parasitism. The success of *C. noacki*, in this case, may stem from the fact that it produces at least three generations per year as opposed to one annual generation of its new host.[32]

In addition to the above, four more whitefly species have been found as hosts for *C. noacki*. One of these, *Aleurotrachelus* near *espunae*, became a preferred host, reaching a parasitization rate of over 30% in some areas.[33] Thus, *C. noacki*, which is the only representative of its genus used for biological control and the only South American representative in our discussion, appears to have an extensive capacity to increase its range to new hosts in new habitats. Such a capacity could prove to be a valuable asset for future biological control of whiteflies.

IV. ADAPTATION OF HETERONOMOUS PARASITOIDS TO NEW HOSTS

A flexible host range was also observed in the selection of hosts for male development of heteronomous parasitoids, i.e., parasitoids with sex-deviate ontogeny. In these species, the female larvae develop as primary parasitoids, but the males show different host relationships, often developing as hyperparasitoids on immature female parasitoids belonging to the same or related species.[34-36]

Since the hyperparasitic development causes the mortality of the primary parasitoid, it may be counteradaptive if the ovipositing female actively selects, oviposits in, and kills female progeny of her own species. Indeed, Williams[35] has shown for *E. tricolor* that a male-producing female will prefer to oviposit in species other than her own. Thus, the host range for male progeny of many *Encarsia* species often extends over a large number of parasitoid species, all of which may be parasitic on the same whitefly host or on a guild of whitefly hosts that occupy similar habitats.

In some cases, the female parasitoid accepts, for male-egg deposition, hosts belonging to distant taxonomic categories. *E. aleuroilicis* Viggiani was found to oviposit male eggs, which successfully developed to adults, in the platygasterid

Amitus vesuvianus Viggiani, a primary parasitoid of *Aleuroviggianus adrianae* Iaccarino, as well as in the ectophagous eulophid *Tamarixia pubescens* (Nees) and several endophagous encyrtids,[37] all primary parasitoids of the psyllid *Trioza ilicinea* (De Stefani-Perez).

Males of a number of heteronomous *Encarsia* species develop in hosts other than parasitic Hymenoptera. The most notable cases are those of male development within the eggs of lepidopterous hosts, first recorded by Flanders[38] and recently reviewed by Polaszek.[36] Since Flanders'[38] report of finding a codling moth egg containing a larva of *Encarsia*, additional cases have been reported both in the U.S.[15, 39] and in Europe, where the males of *E. porteri* Mercet were found to develop in eggs of several Lepidoptera belonging to different families.[40,40a]

V. EVIDENCE FOR RESISTANCE TO HOST-RANGE INCREASE

Gerling[40b] found numerous cases of parasitoid encapsulation within the nymphs of *Trialeurodes vaporariorum* infesting *Sonchus oleraceus* in Hawaii. The species of parasitoids could not be ascertained, but it was probably one of the four species found parasitizing this host in Hawaii at that time.[18] In any case, a defense mechanism, like encapsulation, curtails the success of the parasitoid and therefore constitutes a barrier to range increase by the parasitoid.

Gerling[41] suggested that the great diversity of waxy coverings and protuberances of whitefly immatures might serve for protection from natural enemies. Indeed, a series of recent experiments[41a] with *Aleyrodes singularis* Danzig have shown that the powdery wax secretion, which permanently covers the patches of immature whiteflies in that species, apparently plays a protective role. The parasitoid *E. inaron* that parasitizes this host in nature is able to penetrate the wax covering and reach the invisible whitefly nymphs. However, when *E. transvena* was released on patches of *A. singularis*, the female became entangled in the waxy powder and entirely failed to function. Artificial removal of the wax resulted in successful parasitization of the whiteflies, proving the suitability and susceptibility of unprotected nymphs to parasitoid attack.

VI. DISCUSSION

There are several prerequisites for a full understanding of the characteristics of parasitoid host ranges and their evolution in time. A thorough evolutionary analysis requires knowledge of the exact initial endemic range of the parasitoid and of the full host fauna to which it was exposed. Information is also needed regarding the subsequent movements of the parasitoid and of the new hosts to which it has become exposed. Finally, a list of those new hosts which it attacked and on which it managed to develop successfully is also necessary.

Such complete information is not available for whitefly parasitoids. Moreover, the almost complete absence of reliable biosystematic studies on most whitefly/parasitoid complexes makes the reliability of some of the existing literature records uncertain. Therefore, the "endemic" range often represents a very extensive area which includes numerous geographic and climatic regions, each with its own flora and host fauna. *E. lutea*, for example, has been described from Italy, but it is also known from Spain to Pakistan. Likewise, *E. luteola* is known from both coasts of the U.S. and from many of the southern states, and *E. inaron* (sometimes known as *E. partenopea*) is known from most of the Middle East, including Egypt and Jordan, as well as Central Europe and Britain. Thus, the large number of "original" hosts from which these parasitoids have been recorded may include earlier adaptations to new hosts that might have occurred during their spread from an original, more limited range.

Our data and analyses are based on recent studies and attempts at biological control of whitefly pests, which often serve as an outdoor experiment, examining the degree of parasitoid host-range flexibility. From the host list obtained from this large-scale experiment, it is possible to pinpoint several characteristics concerning host finding and acceptance, host suitability and regulation, and the utilization of these parasitoids for biological whitefly control.

If we consider both the adoption of new hosts in the original habitat and in the new habitat into which the parasitoid was introduced, it is apparent that the increase in host repertoire of each parasitoid was made possible through an increase in the number of different host species available. It would also appear that in many instances the new hosts occupy an ecological niche similar to that of the original ones, e.g., *B. tabaci* and *T. vaporariorum* generally infest the same species of shrubs or annual plants. However, in some cases, there is a significant change in host habitat. *C. noacki*, for example, was brought to Italy as a parasitoid of the tree-infesting *Aleurothrixus floccosus* and has moved to hosts like *Aleyrodes lonicerae* growing on low perennials on the borders of the Mediterranean chaparral. A similar change is noticeable in *E. transvena* in Italy, which moved from the typical host plants of *B. tabaci* to citrus trees on which it found and parasitized *P. myricae*.

The conclusions that can be drawn regarding the mechanisms of host suitability and regulation are very limited, since the details of whitefly physiological and immune reactions have yet to be studied. Host penetration by the parasitoid and immature parasitoid development in the whitefly host have undergone mainly descriptive studies (e.g., Gerling;[15] Gerling et al.;[42,43] Laudonia and Viggiani;[44] etc.). Moreover, although *Cales*, *Encarsia*, and *Eretmocerus* successfully developed in newly acquired host species without evident physiological incompatibility, our observations with *Encarsia inaron* and *E. transvena* indicate that defense, incompatibility, or possible encapsulation may have prevented the successful establishment of new host relationships.

It is interesting to note that host-range increases in the genus *Eretmocerus* have not been noted as frequently or extensively as those of *Cales* or *Encarsia*. For example, *Eretmocerus mundus*, which has as wide a geographic range as *Encarsia lutea*, has not been recorded from as many hosts. The biological significance of this observation may be related to the fact that *Eretmocerus* is an external ovipositor, whose female has to reach the outer margin of the host with the posterior tip of her abdomen in order to insert the ovipositor under the whitefly. Whitefly species that abound in spines or waxy secretions which protrude along their margins may be protected from parasitism by a species that has not specifically coevolved with them.

Biological control theory assumes a certain level of stability in host/parasitoid relationships, at least to a degree in which the introduced parasitoid will attack the pest against which it is intended without becoming itself a pest insect. The introduction of parasitoids into new habitats puts this to the test and can be documented, since the introduction dates and names of the potential hosts are noted. Using the above examples, it is possible to assess the benefits against the risks which are raised by the changes in host range of whitefly parasitoids.

Hokkanen and Pimental[10] found 46 cases of "taxonomic jumps by biocontrol agents that have successfully controlled an evolutionarily new host species." They also claim that such new pest and natural enemy combinations may bring about better biological control than that which would have occurred had the natural enemies been released against their old or original hosts, due to long-term host parasite coevolution, resulting in well-adapted parasites that have little impact on host fitness. Conversely, a new association would have a much greater impact. According to their theory, another benefit of a new association would be its capacity to control native pests, which lack effective enemies. No evidence for the higher fitness theory is available in the cases that we have presented concerning whiteflies. However, the proliferation of some parasitoid species in the new area into which they have been introduced and the adoption of additional hosts in their original area of distribution clearly point to the possible advantage that the host-range increase of whitefly parasitoids can have for the practice of biological control of both introduced and native pests.

Recently, Howarth[45] published a study enumerating the environmental impact of biological control and discussed the possibility that parasitoids, introduced for control of a pest species, might also reduce the populations of harmless insects. Only a few examples of the hundreds of successful biological control cases could be found to fit this category. The best-documented example was that of Hawaiian pentatomids, which included components of the local endemic fauna, that were reduced in numbers following the introduction of parasitoids of *Nezara viridula* (L.). While the danger of whitely parasitoids becoming injurious due to their oligophagy and host-range increases could exist in theory, this possibility should be examined under the practical circumstances which occur in nature. The family Aleyrodidae includes only plant pests of which some are very injurious and others only moderately so. Therefore, parasitoid species that attack only whiteflies, like *Eretmocerus* and *Cales* (save for one possibly aberrant record), should not prove cause for concern.

The heteronomous species of *Encarsia*, whose unmated females attack eggs of Lepidoptera as hosts for male development, merit closer examination. Oviposition in lepidopterous eggs is only an outlet for a small percentage of each female's egg complement, with most eggs being allocated for female production within whitefly hosts. Moreover, the number of *Encarsia* females laying unfertilized eggs is always limited by the number of whitefly hosts for their development. Therefore, the danger to the fauna of the Lepidoptera is minimal. Additionally, most Lepidoptera are also injurious to agriculture. Thus, it would appear that any possibly damaging effect caused by the whitefly parasitoids' ability to increase their host range is outweighed by its potential benefit.

ACKNOWLEDGMENTS

We wish to thank Dr. Alan Bartlett of the Western Cotton Research Laboratory of the U.S. Department of Agriculture in Phoenix, AZ for inviting us to write this chapter. Thanks are also due to Prof. David Rosen of the Department of Entomology, Hebrew University of Jerusalem, for help in the literature survey, and to Ms. N. Paz, Tel Aviv University, for the critical reading of the manuscript.

REFERENCES

1. **Debach, P.,** *Biological Control of Insect Pests and Weeds.* Chapman and Hall, London, 1964.
2. **Vinson, B. S.,** Host selection by insect parasitoids, *Annu. Rev. Entomol.*, 24, 109, 1976.
3. **Vinson, B. S.,** Habitat location, in *Semiochemicals, Their Role in Pest Control,* Nordlund, D. A., Jones, R. L., and Lewis, W. J., Eds., John Wiley & Sons, New York, 1981, 51.
4. **Lawrence, P. O.,** Host-parasite hormonal interactions: an overview, *J. Insect Physiol.*, 32, 295, 1986.
5. **Vinson, S. B. and Dahlman, D. L.,** Physiological relationships between braconid endoparasites and their hosts, the *Microplitis croceipes — Heliothis* spp system, *Southwest. Entomol.*, Suppl. 12, 17, 1989.
6. **Opp, S. B. and Luck, R. F.,** Effects of host size on selected fitness components of *Aphytis melinus* and *A. lignanensis* (Hymenoptera: Aphelinidae), *Ann. Entomol. Soc. Am.*, 79, 700, 1986.
7. **Turlings, T. C. J., Wackers, F. L., Vet, L. E. M., Lewis, W. J., and Tumlinson, J. H.,** Learning of host-finding cues by hymenopterous parasitoids, in *Insect Learning, Ecological and Evolutionary Perspectives*, Papaj, D. R. and Lewis, A. C., Eds., Chapman and Hall, New York, 1992, 51.
8. **Waage, J. K., Carl, K. P., Mills, N. J., and Greathead, D. J.,** Rearing entomophagous insects, in *Handbook of Insect Rearing*, Vol. 1, Elsevier, Amsterdam, 1985, 45.
9. **Hokkanen, H. M. T. and Pimental, D.,** New approach for selecting biological control agents, *Can. Entomol.*, 116, 1109, 1984.
10. **Hokkanen, H. M. T. and Pimental, D.,** New associations in biological control: theory and practice, *Can. Entomol.*, 121, 829, 1989.
11. **Legner, E. F.,** Broadened view of *Muscidifurax* parasites associated with endophilous synanthropic flies and sibling species in the *Spalangia endius* complex, *Proc. Annu. Conf. Calif. Mosq. Vector Control Assoc.*, 51, 47, 1983.
12. **Flanders, S. F.,** Biological control of *Saissetia nigra* (Nietn.) in California, *J. Econ. Entomol.*, 52, 596, 1959.
13. **Russell, L.,** The North American species of whiteflies of the genus *Trialeurodes*, *Misc. Publ. U.S. Dep. Agric.*, 635, 1978.
14. **Arzone, A.,** Ricerche su *Trialeurodes vaporariorum* Westw. (Hem., Hom. Aleyrodidae) ed *Encarsia tricolor* Foerst. (Hym. Aphelinidae) in serra, *Colture Protette*, 5, 45, 1976.
15. **Gerling, D.,** Bionomics of the whitefly parasite complex associated with cotton in southern California, *Ann. Entomol. Soc. Am.*, 60, 1306, 1967.

15a. **Gerling, D.,** personal observations, 1992.

16. **Rose, M. and DeBach, P.,** A native parasite of the bayberry whitefly, *Citrograf*, 67, 272, 1982.
17. **Rose, M. and Rosen, D.,** Biological control of *Parabemisia myricae* (Ruwana) (Homoptera: Aleyrodidae) in California, *Isr. J. Entomol.*, 25–26, 73, 1992.
18. **Gerling, D.,** Observations of the biologies and interrelationships of parasites attacking the greenhouse whitefly *Trialeurodes vaporariorum* (Westw.) in Hawaii, *Proc. Hawaii Entomol. Soc.*, 24, 217, 1983.
19. **Gerling, D.,** Parasitoids attacking *Bemisia tabaci* (Gennadius) (Homoptera, Aleyrodidae) in Eastern Africa, *Entomophaga*, 30, 163, 1985.
19a. **Viggiani, G. and Gerling, D.,** personal observations, 1992.
20. **Pedata, P.A. and Viggiani, G.,** Note su *Encarsia transvena* (Timberlake) parassitoide di Aleirodidi nuovo per l'Italia, *Boll. Lab. Entomol. Agrar. Filippo Silvestri*, 48(1991), 241, 1993.
21. **Viggiani, G. and Mazzone, P.,** Sull'introduzione in Italia di *Encarsia pergandiella* Howard (Hymenoptera: Aphelinidae), parassita di *Trialeurodes vaporariorum* (Westw.) (Homoptera: Aleyrodidae), *Boll. Lab. Entomol. Agrar. Filippo Silvestri*, 37, 39, 1980.
22. **Mazzone, P. and Viggiani, G.,** Nuovi dati sull' *Encarsia pergandiella* (Howard)(Hym.: Aphelinidae), parassita esotico introdotto in Italia contro il *Trialeurodes vaporariorum* (Westw.) (Hom. Aleyrodidae), in *Atti XIV Congr. Naz. Ital. Entomol.*, Palermo-Erice-Bagheria, Tip. Bonfardino, Palermo, Italy, 1985, 855.
23. **Viggiani, G.,** Le specie italiane del genere *Encarsia* Forster (Hymenoptera: Aphelinidae), *Boll. Lab. Entomol. Agrar. Filippo Silvestri*, 44(1987), 121, 1988.
24. **Avilla, J., Viggiani, G., Diaz, X., and Sarasua, M. J.,** Morphological and biological notes on *Encarsia meritoria* Gahan (Hymenoptera, Aphelinidae) a parasitoid of *Trialeurodes vaporariorum* (Westwood) (Homoptera, Aleyrodidae) new in Europe, *Biocontrol Sci. Technol.*, 1, 289, 1991.
25. **Viggiani, G. and Laudonia, S.,** Sulla presenzanza in Italia della *Encarsia meritoria* Gahan. (Hym.: Aphelinidae), parassitoide esotica di Aleirodidi, *Redia*, 74, 135, 1991.
26. **Mound, L. A. and Halsey, S. H.,** *A Systematic Catalogue of the Aleyrodidae (Homoptera) with Host Plant and Natural Enemy Data*, John Wiley & Sons, Chichester, 1978.
27. **Onillon, J. C.,** The use of natural enemies for the biological control of whiteflies, in *Whiteflies: Their Bionomics, Pest Status and Management*, Intercept, Andover, U.K., 1990, 287.
28. **Rose, M. and Wooley, J. M.,** Previously imported parasite may control invading whitefly, *Calif. Agric.*, 2, 4, 1984.
29. **Viggiani, G. and Currado, I.,** Sul parassitismo di *Cales noacki* (How.) (Hym.: Aphelinidae) in uova di *Phalera bucephala* L. (Lep.: Notodontidae), in *Atti XI Congr. Naz. Ital. Entomol.*, Portici-Sorrento, Off. Grafiche, Giannutri, Napoli, Italy, 1978, 317.
30. **Viggiani, G. and Laudonia, S.,** *Aleurotuba jelineki* (Frauenf.) (Homoptera: Aleyrodidae) nuovo ospite di *Cales noacki* Howard (Hymenoptera: Aphelinidae), *Boll. Lab. Entomol. Agrar. Filippo Silvestri*, 41, 139, 1984.
31. **Laudonia, S. and Viggiani, G.,** Osservazioni sulla fenologia e sui parassiti di *Aleurotuba jelineki* (Frauenf.) (Homoptera Aleyrodidae) in Italia, *Boll. Lab. Entomol. Agrar. Filippo Silvestri*, 41, 225, 1984.
32. **Guerrieri, E. and Viggiani, G.,** Osservazioni sull'*Aleurothrixus floccosus* (Mask.) (Homoptera: Aleyrodidae) e sul suo antagonista *Cales noacki* How. (Hymenoptera: Aphelinidae) in Campania, *Ann. Fac. Sci. Agrar. Univ. Studi Napoli Portici*, (Ser. 4)22, 11, 1988.
33. **Liotta, G. and Sinacori, A.,** Parassitismo di *Cales noacki*. How. (Hymenoptera: Aphelinidae) su *Aleurotrachelus* sp., *Nat. Sicil.* (Soc. 4), 10, 21, 1986..
34. **Walter, G. H.,** Divergent male ontogenies in Aphelinidae (Hymenoptera: Chalcidoidea): a simplified classification and a suggested evolutionary sequence, *J. Linn. Soc.*, 19, 63, 1983.
35. **Williams, T. B.,** Sex-ratio strategies in the facultative autoparasitic wasp, *Encarsia tricolor* Foerst, Ph.D. thesis, Imperial College, London, 1981.
36. **Polaszek, A.,** Egg parasitism in Aphelinidae (Hymenoptera: Chalcidoidea) with special reference to *Centodora* and *Encarsia* species, *Bull. Entom Res.*, 81, 97, 1991.
37. **Viggiani, G.,** Note biologiche su *Encarsia aleuroilicis* Viggiani (Hym. Aphlinidae), *Boll. Inst. Entomol. Guido Grandi*, 41, 87, 1986.

38. **Flanders, S. F.,** *Prospaltella* as an egg parasite of the codling moth, *Pan-Pac. Entomol.*, 14, 188, 1925.
39. **Stoner, A. and Butler, G.,** *Encarsia lutea* as a parasite of bollworm and cabbage looper in Arizona cotton, *J. Econ. Entomol.*, 58, 1148, 1965.
40. **Rojas, P. S.,** Nota sobre *Prospaltella porteri* Mercet (Hym.:Aphelinidae) un nuevo parasito de huevos de lepidopteros, *Rev. Chil. Entomol.*, 6, 123, 1968.

40a. **Viggiani, G.,** unpublished data, 1992.

40b. **Gerling, D.,** personal observations, 1980.

41. **Gerling, D.,** Natural enemies of whiteflies: predators and parasitoids, in *Whiteflies, Their Bionomics, Pest Status and Management*, Gerling, D., Ed., Intercept, Andover, U.K., 1990, 147.

41a. **Guerzon, . and Gerling, D.,** in preparation, 1993.

42. **Gerling, D., Orion, T., and Delarea, Y.,** *Eretmocerus* penetration and immature development: a novel approach to overcome host immunity, *Arch. Insect Biochem. Physiol.*, 13, 247, 1990.
43. **Gerling, D., Tremblay, E., and Orion, T.,** Initial stages of the vital capsule formation in the *Eretmocerus-Bemisia tabaci* association, *Redia*, 74, 411, 1991.
44. **Laudonia, S. and Viggiani, G.,** Osservazioni sugli stadi preimaginali di *Cales noacki* Howard (Hym: Aphelinidae), *Boll. Lab. Entomol. Agrar. Filippo Silvestri Portici*, 43, 21, 1986.
45. **Howarth, F. G.,** Environmental impacts of classical biological control, *Annu. Rev. Entomol.*, 39, 485, 1991.

Chapter 10

INCREASING GENETIC DIVERSITY FOR RELEASE OF PARASITES AND PREDATORS

Fred D. Bennett

TABLE OF CONTENTS

0-8493-2607-9/94/$0.00+$.50

I. INTRODUCTION

The goal of classical biological control is the field release of effective biotic agents for the permanent suppression of the target pest without causing any adverse impact to the ecosystem. The standard biological control protocol includes exploration, host-range testing, and quarantine screening. Often, only a small proportion of the genetic diversity present in the area of origin is represented in consignments sent to quarantine. This is diminished further during testing and screening, and the final laboratory colony may contain very limited genetic variation by the time of field release. Except where the target pest is clearly not a suitable host for the development of a natural enemy or there are climatic factors which do not permit the natural enemy to survive, we cannot account for lack of establishment or the poor performance of an introduced natural enemy. These and related aspects have been discussed by DeBach,[1] Doutt and DeBach,[2] and Messenger, Wilson et al.[3]

In this chapter, I do not propose to discuss genetic improvement. However, some of the examples which I will mention involve the cross-mating of different geographic isolates to increase genetic diversity. This could be considered genetic improvement. To clarify this, I quote from Hoy:[4] "*Identification* and use of naturally occurring biotypes is not genetic improvement. Nor is the maintainance of genetic quality considered genetic improvement, although the *maintainance* of quality must be an integral part of genetic improvement."

Once genetic improvement has been incorporated into a laboratory population, the steps required to maintain the genetic quality of the improved strain are similar to those required in retaining the genetic diversity present in the biotypes introduced for classical biological control. In fact, it may be easier to gauge the former because it is possible to measure the persistence, or disappearance of the quality or trait (e.g., level of pesticide tolerance), originally achieved by genetic improvement. It is much more difficult to determine whether diversity has diminished in a natural strain when no parameters to measure fitness have been established.

I have restricted my presentation to examples, mainly in the neotropics, where repeated introductions from the same or different areas, or attempted laboratory improvement, appear to have led to the effective establishment of a natural enemy or to an improved level of biological control by one already present.

II. NUMBERS OF INTRODUCED NATURAL ENEMIES

Do we need to be concerned about the size or numbers of individuals involved in the introduction of a natural enemy introduced to a new environment? In many cases, apparently "NO!".

As a prime example, consider the establishment and spread of the vedalia beetle *Rodolia cardinalis* and its success in California from one introduction. Success has been repeated many times following introduction of the vedalia beetle from California into other countries. Often, only a relatively small genetic sample was represented because only very small subsets of the original California introduction were included.

Many, if not most, of the highly successful biological control projects have occurred from rather limited introductions and from what would appear to be a very limited genetic base. Therefore, there are other factors, such as exclusion of pathogens, which might take priority over concern about genetic diversity during the protocol of clearing exotic natural enemies for field release.

On the other side of the coin, remember the failure, or only partial success, of the vedalia beetle in coastal California and Bermuda. Was genetic diversity too limited to perform adequately under cooler climatic conditions? Are there biotypes somewhere which would perform better under cooler climates?

III. NEOTROPICAL EXAMPLES

I will deal with examples which include the dramatic build-up of the Amazon fly *Metagonastylum minense* (Tns.) in St.Lucia and Venezuela, the rapid establishment of the Cuban fly *Lixophaga diatraeae* in St. Kitts, and the acclaimed success of *Cotesia flavipes* in northeast Brazil. Each of these occurred as the result of a single, or at most a series, of limited introductions. I will mention briefly Box's[5,6] attempts to improve the perfomance of the tachinid *Paratheresia claripalpis*, a parasite of *Diatraea* spp., in Venezuela. Also, I will examine the failure, for many years, to get establishment of the tachinid fly *L. diatraeae* in Barbados and the initial failure of *C. flavipes* in the state of Sao Paulo, Brazil; in certain Caribbean islands; and in Venezuela.

A. *METAGONASTYLUM MINENSE* IN ST. LUCIA

The Amazon fly *M. minense*, which had been successfully introduced from Brazil into Guyana for the control of sugarcane borers, was sent to St. Lucia in 1934. From a consignment of 220 puparia, 79 mated females were obtained. Thirty of these were retained to initiate a laboratory colony, and the remaining flies were released in cane fields infested with sugarcane borer *Diatraea saccharalis*. Several of the sugarcane borer larvae collected a few days later to serve as hosts for the laboratory colony had already been parasitized by the released females. *M. minense* established and spread so rapidly that the breeding program was abandoned. Damage, measured in percentage of bored internodes, dropped from 17.2% in 1934 to 1935 to 4.3% in 1938 in the valley where the initial release was made. The parasite spread rapidly throughout the sugar production areas of St. Lucia and continued to provide satisfactory control of *D. saccharalis*.[5] Hence, from a relatively restricted genetic base, rapid establishment and host suppression occurred.

Would the same level of success have been achieved if, as required under today's quarantine restrictions, laboratory breeding had been a requirement before permission to make field releases?

B. *PARATHERESIA CLARIPALPIS* IN VENEZUELA

During an extended campign to control the complex of sugarcane stalk borers in Venezuela, Box[6] imported genetic material of the tachinid parasite of *P. claripalpis* from Trinidad, Mexico, and Peru and compared their rate of successful parasitism with the Venezuelan race when inoculated onto larvae of the four most serious pest

TABLE 1
Levels of Parasitism Obtained in the Laboratory from Inoculations with Different Strains of *Paratheresia claripalpis* on Different Species of *Diatraea*[a]

Race of parasite	Species of *Diatraea* inoculated			
	saccharalis	*rosa*	*busckell*	*lineolata*
P. Venez. (P V)	++	+	(+)	+++
P. Trinidad (P T)	+++	++	....	+
P. Venez. × Trin. (P V × T)	++	++	++	++
P. Mexico (P M)	+++	+++	++	++
P. Mex. × Trin. (P M × T)	++	+++	+	+
P. Peru (P P)	+++	+++	+++	+

Note: (+) = less than 6.5%; + = low parasitism 6.5–13%; ++ = medium parasitism 24–43.5%; and +++ = high parasitism 49% or higher.

[a] Modified from Box, H. E., *Biol. Tec. Inst. Nac. Agric. Maracay*, 5, 1, 1952.

species of *Diatraea*. He also conducted some "hybridization" experiments (Table 1) and concluded that the Trinidad × Venezuelan hybrid appeared to be the best combination for general conditions in Venezuela. Some field releases were followed by "unprecedented increases in natural populations," but Box[6] concluded that proof of establishment would be difficult to obtain because of the absence of any morphological differences in the adults or larvae. Whether or not these experiments might be considered genetic improvement, they do demonstrate that there are well-defined biological differences between geographic populations of *P. claripalpis*. Therefore, initial failure with the introduction of one geographic population should not preclude other trials which incorporate greater genetic diversity.

C. *LIXOPHAGA DIATRAEAE* IN BARBADOS

The Cuban fly *L. diatraeae*, when introduced into St. Kitts in 1932, established rapidly and provided control without requiring annual releases. Based on this success, the Cuban fly was introduced into several other islands in the Caribbean with varying levels of success.[7] In Barbados, despite repeated introductions and the release of several thousand parasites during a 25-year period (1933 to 1958), establishment was not achieved.

In 1958, a program to determine once and for all whether this parasite could be established was initiated. To provide for greater genetic diversity (an expanded gene pool), cultures were obtained from Jamaica, Cuba, and the Dominican Republic, where this parasite occurs naturally, and from St. Kitts and Antigua, where it had been introduced and provided some level of control of *D. saccharalis*. These different geographic races were cross-mated in the laboratory, and over 500,000 adults were released over a 5-year period. Only sporadic recoveries, usually shortly after a release, were reported with no evidence of parasite buildup or permanent establishment.

After the main mass-breeding and release program was curtailed and attention was focused on the introduction of other natural enemies, limited releases of the Cuban fly were continued. As opportunity permitted additional genetic material, usually in the form of adults reared from field-collected puparia and field-parasitized host larvae obtained during surveys in St. Kitts and Antigua, were either released directly into the field or added to the laboratory colonies for future releases.

As sporadic, often seasonal, pockets of establishment occurred, I suggested that a laboratory colony based on adults emerging from field recoveries in Barbados should be set up to form the basis of future field releases. The rationale behind this suggestion was based on the possibility that these flies represented progeny from individuals better adapted to Barbados conditions. It was only after this program was initiated that *L. diatraeae* was recovered consistently and became abundant enough to have an economic impact on sugarcane borer populations. Even today, we do not know whether permanent establishment occurred as a result of our attempts to select a population adapted to Barbados or as a result of small releases of field-collected material from Antigua or St. Kitts during the later stages of the program.

Today, it should be possible by comparative isoenzyme analysis, DNA analysis, or other modern techniques to ascertain whether the Barbados population matches one of the parent populations from Cuba, the Dominican Republic, or Jamaica or whether it represents a hybrid of two or more of the geographic races.

D. *COTESIA FLAVIPES* IN BARBADOS

It is ironic and at the same time provides a striking contrast between this long, drawn-out campaign and the ease that another parasitoid became established in Barbados just at the time that the Cuban fly started to take hold. The braconid *C. flavipes* was shipped to Barbados as one of several natural enemies of allied Asian stemborers for trial after it appeared that the campaign to establish the Cuban fly was unlikely to succeed. In fact, Van Whervin,[8] after detailed ecological studies, had concluded that Barbados was "environmentally resistant" to the Cuban fly.

Two thousand adults of *C. flavipes* emerging from cocoons sent from India were released in a single field during July and August 1966. No further action was taken until this braconid was recovered in October 1967, at which time a mass-breeding and release program was initiated. This resulted in island-wide establishment, and *C. flavipes*, in combination with the Cuban fly, has provided continuous biological control.[9] Hence, with an importation of limited genetic diversity, *C. flavipes* became established and aided in the successful control of *D. saccharalis*, whereas the Cuban fly only became established after a campaign where a concerted effort was made to broaden genetic diversity.

E. *COTESIA FLAVIPES* IN BRAZIL

Following the success of *C. flavipes* in Barbados, requests for this parasite came from several other countries. The Commonwealth Institute of Biological Control (CIBC) in Trinidad supplied breeding stocks of the Indian race to Brazil to initiate a control program in northeast Brazil and later in Sao Paulo. In northeast Brazil, the braconid introduced in 1974 took hold immediately, and as a result of an ambitious mass-breeding and release program, it rapidly provided excellent control.[10] Macedo et al.[11] calculated that the cost of $430,000 (U.S. dollars) provided benefits of $111 million accrued over a 10-year period. By contrast in Sao Paulo, despite large-scale releases, levels of parasitism remained low, and the releases did not provide satisfactory control. I recommended trials with a cool-weather strain from Pakistan. After the introduction of this race in 1978, the picture changed dramatically. Following releases, levels of field parasitism rose dramatically, and by the end of 1984 more than 25 laboratories had been built to provide parasites for the sugar industry, which, at this time, was expanding rapidly to produce alcohol as a replacement for gasoline.[10] Parasite production exceeded 1 billion per annum during 1985.[12]

F. *COTESIA FLAVIPES* IN GUADELOUPE

In Guadeloupe, initial efforts to establish *C. flavipes* were unsuccessful despite repeated releases of material obtained from Barbados. In 1989, new cultures were obtained from Barbados and also from Martinique, where establishment had occurred. These populations, both originally of Indian origin, were hybridized and released without prolonged laboratory breeding. In contrast to the earlier attempts, establishment and significant levels of parasitism occurred rapidly.[13]

G. *COTESIA FLAVIPES* IN VENEZUELA

C. flavipes was introduced into Venezuela from Trinidad in 1981. Although initial parasitism as high as 35% was observed, effective establishment did not occur, and this species was not an important control agent.[14] In 1987, *C. flavipes* was reared from *Diatraea* sp. found in Tachira State, Venezuela, near the Colombian border. A laboratory colony was established, large-scale releases were made in Portugese and Yaracuy states, and effective control occurred. According to Ferrer, Proano, and Salazar,[14] *C. flavipes* has almost completely displaced the Amazon fly, and, perhaps more important, it successfully attacks *D. centrella*, which was seldom successfully parasitized by the Amazon fly.

The source of this effective strain in Venezuela is not clearly understood. It may have developed from survivors of the 1981 releases in Venezuela or it may have migrated from nearby sugarcane areas in Colombia where in previous years large-scale releases had been made. The Colombian biological control program included several introductions of both the Indian and Pakistan races of *C. flavipes*, and interbreeding may have resulted in the selection of a locally adapted strain.

IV. CONCLUSIONS

Although the introduction of a limited number of *L. diatraeae* into St. Kitts, *M. minense* into St. Lucia, and *C. flavipes* into Barbados provided the required genetic base in the relatively small founding colonies to effect efficient control, this did not occur with several of the other introductions. The various examples that I have discussed point to the value of persistent effort, and in instances where initial introductions fail, to obtain additional material from the same or other areas.

Based on experience with weed biocontrol agents in the Gainesville, FL quarantine facility, a significant loss in genetic diversity may occur during the prolonged time period required for host-range testing and the permitting process before field releases are authorized. Two recent instances suggest that during this interval genetic diversity diminished to the extent that field establishment did not occur.

The initial releases of *Hydrellia pakistanae* introduced into Florida for control of hydrilla, comprised of material bred in the laboratory for several generations, failed in the field. A new importation released as F_1 and F_2 established immediately.[14a]

The same sequence of events occurred with *Spodoptera* (= *Namangana*) *pecticornis* introduced from Thailand for the control of water lettuce. Despite large-scale releases from a colony reared in quarantine for many generations, establishment did not occur. In contrast, a recent release (1992) based on progeny of a new introduction from the same source in Thailand resulted in impressive initial damage on a scale not seen during the earlier introductions. Unfortunately, the release site was treated with herbicides which destroyed the water lettuce and precluded further observations.[14b]

It is frequently argued, but seldom convincingly demonstrated, that a decrease in the genetic diversity is the major cause in the poor performance of an introduced natural enemy. Nevertheless, as this may account for the failure of establishment or may explain why certain control agents are only partially successful, care should be taken to preserve genetic diversity during quarantine breeding or to incorporate new genetic material into the cultures before large-scale breeding and field releases are initiated.

Baker[15] drew attention to the occurrence and differences in strains of *C. flavipes*. In areas where control of sugarcane borers has not been adequate, Baker[15] recommends trials with other strains.

In conclusion, I have attempted to use examples which indicate that attempts to broaden the genetic base, in other words to increase genetic diversity, have resulted in improved levels of control. Therefore, if initial releases have failed, it is advisable to make other importations in order to incorporate additional genetic material into the material being reared for field release.

Bennett,[16] when reviewing biological control operations in the Caribbean, concluded that many of the "failures" in biological control were due to insufficient funding, with the result that there were insufficient resources and an insufficient breadth of genetic material introduced to ensure a reasonable chance of success. The same conclusions apply to most areas of the world. If it is not possible to maintain genetic diversity indefinitely in laboratory cultures, then new genetic material should be introduced if initial field releases are not successful.

ACKNOWLEDGMENTS

I wish to thank Dr. Marjorie Hoy for providing me with relevent background references.

REFERENCES

1. **Debach, P.,** Selective breeding to improve adaptations in parasitic insects, *Proc. 10th Int. Congr. Entomol. Montreal*, 4, 759, 1958.
2. **Doutt, R. L. and Debach P.,** Some biological control concepts and questions, in *Biological Control of Insect Pests and Weeds*, Debach, P., Ed., Chapman and Hall, London, 1964, 118.
3. **Messenger, P. S., Wilson, F., and Whitten, M. J.,** Variation, fitness, and ability of natural enemies, in *Theory and Practice of Biological Control*, Huffaker, C. B. and Messenger, P. S., Eds., Academic Press, New York, 1976, 209.
4. **Hoy, M. A.,** Genetic improvement of arthropod natural enemies: becoming a conventional tactic?, in *New Directions in Biological Control: UCLA Symposium. Molecular & Cellular Biology*, Vol. 112, Baker, R. and Dunn, P., Eds., Alan R. Liss, NewYork, 1990, 405.
5. **Box, H. E.,** Investigaciones sobre los taladradores de la cana de azucar *Diatraea* spp. en Venezuela. El proyecto del combate biologico: informe del progreso durante 1949–1951, *Bol. Tec. Inst. Nac. Agric. Maracay*, 5, 1, 1952.
6. **Box, H. E.,** Status of the moth-borer *Diatraea saccharalis* (F.) and its parasites in St.Kitts, Antigua and St. Lucia; with observations on Guadeloupe and an account of the situation in Haiti, *Proc. Int. Sugarcane Technol.*, 10, 901, 1960.
7. **Bennett, F. D., Cock, M. J. W., Hughes, I. W., Simmonds, F. J., and Yaseen, M.,** Review of biological control in the Commonwealth Caribbean and Bermuda up to 1982, Cock, M. J. W., Ed., *Commonw. Inst. Biol. Contr. Tech. Commun.*, 9, 1, 1985.
8. **Van Whervin, L. W.,** Ecology, biology and behaviour of *Lixophaga diatraeae* Tns. and results of attempts (1958–1963) to establish it in Barbados for the control of moth borer, *Diatraea saccharalis* (F.), *Bull. Sci. Agric. Barbados*, 43, 1, 1966.
9. **Alam, M. M.,** Biological and ecological factors affecting the control of sugarcane moth-borer *Diatraea saccharalis* (Lep.: Pyralidae) Barbados, W.I., *Entomophaga*, 25, 401–414, 1985.
10. **Bennett, F. D.,** Biological control of sugarcane borers *Diatraea* spp. in Latin America, *Proc. 1985 Meet. West Indies Sug. Technol. Trinidad and Tobago*, 2, 496, 1985.
11. **Macedo, N., Mendoca, A. F., Moreno, J. A., and Pinazza, A. H.,** Evaluation of the economic advantages of 10 years of biological control of *Diatraea* spp. through *Apanteles flavipes* (Cameron), in the State of Algoas (Brazil), *Int. Sugarcane Technol. Entomol. Newslett.*, 16, 9, 1984.
12. **Macedo, N. and Botelho, P.,** Ten years of biological control of *Diatraea saccharalis* by *Apanteles flavipes* in Sao Paulo state (Brazil), *Proc. Int. Sugarcane Technol. Congr.*, 19, 551, 1986.
13. **Cochereau, P.,** Installation en Guadeloupe de *Cotesia flavipes* Cameron (Hymenoptera: Braconidae), un parasite larvaire du foreur de la canne a sucre *Diatraea saccharalis* L. (Lepidoptera: Crambidae), in *Rencontres Caraibes en Lutte Biologique*, No. 58, Pavis C. and Kemarrec, A., Eds., Les Colloques d l'INRA, Paris, 1991, 443.
14. **Ferrer, F., Proano, L., and Salazar, J.,** El effecto de control biologico de *Cotesia flavipes* (Hym. Braconidae) sobre *Diatraea* spp. (Lep. Pyralidae) en el area de influencia de la azucarera Rio Turbio, durante los años 1986-1990, *Venezuela Azucarera*, 36, 25, 1990.

14a. **Buckingham, G.,** personal communication, 1992.

14b. **Habeck, D. H.,** personal communication, 1992.

15. **Baker, P. S.,** Biocontrol in the Caribbean by IIBC: past, present and future trends, in *Rencontres Caraibes en Lutte Biologique*, No. 58, Pavis, C. and Kermarrec, A., Eds., Les Colloques d l'INRA, Paris, 1991, 63.
16. **Bennett, F. D.,** An overview of classical biological control in the Caribbean and some examples of the utilization of entomophagous insects, in *Rencontres Caraibes en Lutte Biologique*, No. 58, Pavis, C. and Kermarrec, A., Eds., Les Colloques d l'INRA, Paris, 1991, 25.

Chapter 11

IMPROVEMENT OF NATURAL ENEMIES FOR BIOLOGICAL CONTROL: A GENETIC ENGINEERING APPROACH

L. J. Heilmann, J. D. DeVault, R. L. Leopold, and S. K. Narang

TABLE OF CONTENTS

0-8493-2607-9/94/$0.00+$.50

I. INTRODUCTION

Biotechnology has tremendous potential in the area of insect control. This could range from the development of genetic sexing strains, the engineering of disease resistance for beneficial insects, the control of insect transmitted diseases to improving the effectiveness of natural enemies against insect pests.[1-3] The successful use of genetic engineering for the improvement of insect parsitoids is a multistage project which requires the integration of a number of distinct technologies (Figure 1). Initially, traits that are desirable must be identified and characterized. Next, the gene(s) involved in the expression of the desired trait must be cloned and characterized. Finally, these genes must be suitably modified and successfully introduced into the target organism's DNA such that the desired trait will be transmitted in a Mendelian fashion.

The aim of this chapter is to discuss ways in which genetic technology could benefit the use of insect parasitoids in pest control. We will first discuss the general concept of genetic manipulation and how it applies to biological control. Next, we will analyze those traits which may prove to be useful in biological control and discuss the various techniques available for identifying and isolating the genes involved. We will then turn our focus toward a description of the methodologies currently being used to introduce cloned DNA into target organisms. Finally, we will suggest future research directions and identify potential problems that will need to be addressed in order to achieve maximum potential from the use of genetic technology on insect parasitoid biocontrol.

II. GENETIC MODIFICATION: AN OVERVIEW

In a general sense, genetic modification is the ability to manipulate the genome of a living organism in such a way as to produce a desired effect. Therefore, prior to manipulating the genome of a parasitoid, some characterization of the genome will have to be accomplished.[4] Mapping of both naturally occurring and induced mutations would provide markers for genetic transformation, as would other mapping methods such as restriction fragment length polymorphisms (RFLPs),[5,6] microsatellite sequences,[7] and random amplified polymorphic DNA polymerase chain reaction (RAPD-PCR).[8] There are as many strengths as there are weaknesses for each mapping method; therefore, no one method is the method of choice. The required end result is a physical genome map, which will provide genomic landmarks which will be useful later for identifying the sites of recombinant DNA insertion.

The question of which traits or behaviors to modify is very important. A number of potentially useful traits will be discussed in greater detail in Section III and are summarized in Table 1. Generally speaking, the trait should be monogenic and dependent upon the action of a single gene, since only a single gene will be added or changed within the genome. This criterion eliminates many complex behaviors which are polygenic in nature and beyond the capabilities of present technology. For the most part, pesticide resistance represents the only monogenic trait listed in Table 1. Resistance to pathogens and environmental fitness traits such as cold hardiness or

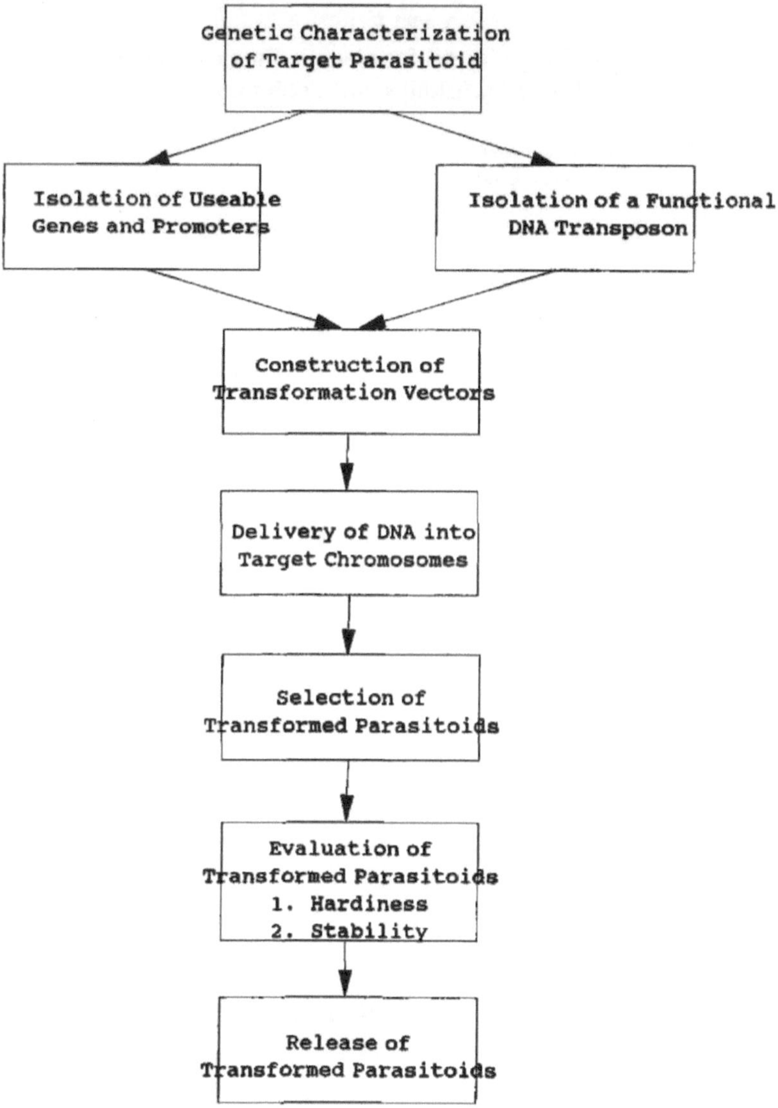

FIGURE 1. Diagram of the steps necessary to genetically transform insect parasitoids.

the ability to diapause do have monogenic components; however, these usually only modify the trait, not change it. Traits such as host selection are very complex multigenic behaviors. While many of these behavioral traits might seem beyond reach, it has been shown in *Drosophila* that behaviors as supposedly complex as circadian rhythmicity[9] and memory[10] are open to genetic correction or modification. The traits to be modified must also be regulated by genes that are clonable. While homologous genes from the target organism are preferable for cloning, often this is not possible. In some instances, genes and gene promoters from other organisms can

TABLE 1
Potential Target Traits for Genetic Engineering in Insect Parasitoids and Predators

1.	Pesticide resistance
2.	Pathogen resistance
3.	Diapause control
4.	Environmental hardiness (high- and low-temperature tolerance)
5.	Sex ratio
6.	Host-seeking ability
7.	Host preference range

be substituted. A number of pesticide resistance genes have now been cloned,[11,12] as well as some diapause-specific genes.[13,14] Whether these genes would work in heterologous systems remains to be determined.

Development of a method to genetically transform or insert the modified gene into the chromosomes of the target parasitoid is the key technology for the entire project. At present, the only operational transformation system for insects is the *Drosophila* P-element system,[15] and it has been shown to be very species specific. The success of parasitoid genome manipulation requires both a transformation method as well as a transformation vector. The transformation vectors must incorporate a dominant selection system to eliminate those insects which have not incorporated the new DNA. We discuss at length in Sections V to VII a number of different possible methods for introducing the DNA into embryos and for finding or designing vector DNAs to properly package the gene before insertion.

Once transgenic strains of the parasitoid have been created in the laboratory, they must be evaluated for stability under mass-rearing conditions. This must also include an assessment of their relative fitness and competitiveness with wild-type parasitoids in the field as well as the functionality of the modified trait. The stability of the newly inserted DNA must also be evaluated for its expression as well as its functionality.

III. GENETIC BASIS OF DESIRABLE QUALITATIVE AND QUANTITATIVE TRAITS

Traits or genetic markers such as allozymes, pesticide resistance, and visible phenotypic mutations, where one or a few genes are involved, follow simple Mendelian inheritance patterns. In contrast, many adaptive traits are polygenic. The expression of such quantitatively inherited traits depends upon the effect of variable environments and the genetic differences at various loci affecting the trait.

Although, for the most part, hymenopteran species, including those used in biological control, generally display lower levels of electrophoretic variations than other insects[16] (see Chapter 2), several adaptive physiological and behavioral traits have been identified which can be selected for improvement.[17] Some desirable traits, such as pesticide resistance, can be easily selected for by traditional selection methods. Strains of the predatory mite resistant to multiple pesticide types were created by selective breeding in the laboratory[18] and were released in areas subject to pesticide applications. Recently, two strains of a predatory mite, *Typhlodromus pyri*, resistant to synthetic pyrethroid insecticides have been bred by selective breeding programs (see Chapter 6). Laboratory selection for pesticide resistance in natural enemies has been discussed in Chapter 5.

In spite of considerable research effort during the past five decades, reproducible methods of selection for quantitatively inherited traits (QTL) remain elusive. There are no reliable methods to determine (1) the number of loci controlling variation; (2) the relative effects of alleles of different loci; (3) additive, dominant, pleiotropic effects and epistatic interactions among them; and (4) the mutation rate. In addition, selection is unpredictable due to a lack of adequate understanding of the genes controlling the response to short- and long-range natural and artificial selection.[19]

It is beyond the scope of this chapter to describe concepts and methodologies to determine genetic basis of QTLs or to predict their response to selection and maintenance, including long-term evolution of genetic variance (which depends on the distribution of allelic effects of individual loci). Some discussion on statistics such as the number of genes and their alleles, phenotypic values vs. their frequencies, covariance, correlation coefficient and intraclass correlation coefficient among traits, and realized heritability and artificial breeding or selective breeding (response to selection, change in allele frequency, selection differential, etc.) can be found in Chapter 5.

IV. SELECTABLE TRAITS

We will concentrate on three broad areas which could be open for genetic engineering work aimed at improving the effectiveness and economic usefulness of insect parasitoids (Table 1). These are (1) resistance to pesticides and pathogens, which would improve the parasitoids utility in the field and ease problems in mass rearing; (2) environmental hardiness to insure that the released parasitoids can function and reproduce in their new environment for extended periods of time after the initial release; and (3) sex-ratio distortion, which will allow manipulation of populations both in the laboratory and the field and which could ease problems of artificially rearing large enough populations for release. These are traits which are either monogenic or have at least some monogenic characters or components that can be selected for. Two other traits, host-seeking ability and host preference, are polygenic in character and will be much more difficult to modify.

A. PESTICIDE RESISTANCE

Pesticide resistance is one of the persistent problems of entomology. It is a classic example of evolution in action. Heavy pesticide use has selected for the presence of certain genes that confer resistance or increased tolerance to individual pesticides or families of pesticides. Resistance has been detected in hundreds of different species of insects. In most cases, resistance is inherited in a simple Mendelian fashion indicating that it is the result of a single-gene mutation.

The genes responsible for resistance to a number of classes of pesticide have now been cloned and are under study[11,12] (see Chapter 6). Based on their mechanisms of resistance, they can be divided into three main categories: (1) genes coding for proteins that metabolize the pesticide molecule once it gets into the body (cytochrome P450, glutathione transferases, and esterases),[20-22] (2) genes coding for pesticide targets such as neurotransmitters[23] or ion channel proteins,[12] and (3) genes coding for proteins that effect entry of the pesticide into the insect.[24]

Techniques are now at hand to turn the problem of pesticide resistance into a useful trait.[1-6] It is difficult to use parasitoids in many areas and with some crops because subsequent or even previous pesticide use can destroy the parasitoid as effectively as the targeted insect. One method around this problem would be to use parasitoids that are resistant to the pesticide in use. This would, theoretically, allow not only the use of the parasitoid, but also its use in planned combination with the pesticide, increasing the effectiveness of both. Extensive work has been done attempting to artificially select for resistant insects in laboratory strains of many species.[25] Laborious selection has resulted in some strains which are measurably more resistant, but not resistant enough, to be useful in the field.[26] This is one avenue which may be easily exploited by using biotechnological techniques.

The basic approach would be as follows. A gene encoding a pesticide resistance determinant (i.e., cytochrome P450) would be modified so that its translated product would be of a high functional quality. The gene would then be introduced into the genome of the parasitoid. Once the gene is stably expressed, the parasitoid will be more resistant to the particular pesticide that the protein metabolizes. Use of this pesticide could then continue after release of the parasitoids without fear that application of the specific pesticide would have a negative effect on the parasitoid.

B. DISEASE RESISTANCE

Disease is a common problem in any mass-rearing situation and can affect competitiveness and survival in the wild after release. These diseases can be the result of bacterial or protozoan infections or even other parasitoids such as mites which infect hymenopterans. Insects do not fight pathogens with mammalian-type immunoglobulins, but they do have several systems of defense which have recently received considerable attention.[27] In some insects, bacterial infection triggers the production of several different families of proteins having bacteriocidal activity.[28] In addition, some larger organisms are encapsulated, a principle mechanism many insect hosts use to escape parasitoids[27]. If such a gene could be added to a parasitoid under the proper control, it might have an enhanced ability to fight off infections or parasites. Consequently, the lifespan of the parasitoid and, hence, its reproductive output would be increased. Likewise, manipulation of genes for defensive proteins like cecropin or defensin[29] or addition of new defensive genes could enhance the parasitoid's response to microbial challenge.

C. ENVIRONMENTAL HARDINESS

A major problem with parasitoid release is reduced survival or nonreproduction of the released parasitoid. Often, the climate or other aspects of the environment are not optimal. This is particularly a problem with exotic parasitoids introduced into an area far from their normal range. For example, tropically adapted insects will probably not overwinter in temperate climates. While environmental hardiness is a polygenic trait, certain components have been identified that could be exploited. Considerable work is being conducted on just how insects manage to survive winter temperatures.[30] If individual genes responsible for these traits can be identified, they can then be isolated and transferred to insects that one might wish to make more cold hardy. A possible example is the gene(s) responsible for the synthesis of glycerol, a polyol that is used by some insects as a sort of antifreeze.[31]

Diapause is another mechanism used by many insects to survive hostile seasonal weather. While there appear to be a number of different types or levels of diapause, the actual mechanisms of induction and control remain mysterious. Diapause and nondiapause traits can be selected for and genetically modified. Hoy[32] selected a nondiapausing strain of gypsy moth, which was used in the laboratory for continuous mass rearing of parasitoids. Proteins that are diapause specific have been identified,[33] as well as a diapause-inducing polypeptide.[14] Further work will eventually lead to the controlling genes. If the mechanisms governing this response in the parasitoids are similar, it should be possible to use these genes and those for diapause-inducing hormones to selectively induce diapause in the parasitoids. It is also very possible that in some situations one would want just the opposite — parasitoids that do not survive or reproduce in the wild. It might be desirable from an environmental point of view not to permanently add an exotic species to the local fauna. It should be possible to add genes to an insect that will make it more susceptible to some component of the local environment. Likewise, antisense genes[34] could be used to turn off some vital function. These genes could be attached to regulatory sequences which would be activated only under certain desirable conditions and would function as dominant conditional lethals.

D. SEX RATIO MANIPULATION

In many mass-rearing situations, it would be desirable to have an overabundance of one sex, usually females, since they parasitize the host. Strong bias toward females will result in a more efficient production of parasitoids which will kill the target. Sex is determined in most hymenopteran parasitoids by haplodiploidy, with the females being diploid and the males haploid.[35] The males develop from unfertilized eggs. The ratio of males to females in these cases is regulated by the percentage of eggs produced by the female which are fertilized. The female regulates this by controlling the release of sperm stored after mating.[36] Mutations that lead to changes in this ratio would be useful in leading to the genes that might regulate this female behavior. Modification of these genes could then be used to regulate the male/female ratio by controlling the rate of fertilization.

The sex ratio in a number of parasitoids and other hymenopterans is often regulated by such things as temperature[37] or host size.[38] Chromosomal changes have also been found that affect the ratio.[39] This implies that, at least in some species, the ratio is genetically controlled. Modifying the genes controlling this trait could provide the ability to change the ratio of males to females in a rearing situation or even in the wild. While the genetic mechanism of sex determination in hymenopterans is under much debate, many sex-determining genes of the fruit fly *Drosophila*

TABLE 2
Approaches for Integration of Cloned Genes into the Insect Genome

1. Direct DNA injection
2. Injection of transposable element vectors
3. Biolistics
4. Electroporation
5. Virus infection
6. Pole cell transfection
7. Sperm transformation

melanogaster have been cloned and characterized.[40] Either these genes, sex lethal, transformer, and doublesex, or their hymenopteran equivalents would be excellent candidates for use in creating engineered parasitoids.

The sex ratio of hymenopteran parasitoids has also been found to be regulated by so-called sex-ratio distorting factors.[41] These can have multiple origins. A chromosomal factor in *Nasonia vitripennis* called paternal sex ratio (psr) causes all offspring of males carrying it to be males.[42] Two other factors, also in *Nasonia*, have very different sources. Females carrying maternal sex ratio (msr) produce male offspring only when they are virgins and only females after mating.[43] The msr factor is maternally inherited, suggesting that it is an extrachromosomal element.[43] The son-killer (sk) trait is caused by infection with a bacterium.[44] Infected females produce mostly female broods due to most of the male offspring dying as larvae. Details of how the sex-ratio distortion occurs is not known. This is obviously a fertile area for genetic and molecular biological investigation.

V. GENETIC TRANSFORMATION: APPROACHES AND METHODOLOGIES

Once the traits for genetic engineering have been characterized, and the corresponding genes cloned, the next step involves introducing the gene of interest into the new host genome. Genetic transformation is a relatively new concept in entomology and has been well demonstrated only in *Drosophila*.[15] The key step in this process is the insertion of the modified or artificial gene into the chromosomal DNA of the insect in such a way that it is not only expressed properly, but is stably inherited by future generations. The only insect for which this is currently possible is *D. melanogaster*. Research is under way to develop similar systems in other more economically and medically important insects. To date, little or no progress has been achieved with parasitic wasps. However, Presnail and Hoy[45] have reported transforming a parasitic mite by direct injection of DNA. Whether a similar approach may be taken with parasitic wasps has yet to be determined. A major reason that *Drosophila* was the first insect for which transformation was achieved is the abundance of genetic information available about this organism. Methods that have worked in other systems and which might work in parasitoids are listed in Table 2.

TABLE 3
Insects Transfected as Embryos by Microinjection

Species	Ref.
Diptera	
Drosophila melanogaster	15
Anastrepha suspensa	64
Ceratitis capitata	96
Anopheles gambiae	49
Aedes triseriatus	50
Lucillia cuprina	97
Musca domestica	135
Lepidoptera	
Bombyx mori	98,52
Orthoptera	
Locusta migratoria	99

VI. DNA DELIVERY METHODS

A. EMBRYO INJECTION

The effectiveness of any genetic transformation system relies in part on the method whereby the DNA construct is to be introduced into the organism. The method of choice with insects has been microinjection of the embryo at the preblastoderm stage. This method involves chemical or mechanical removal of the egg chorion, desiccating the dechorionated eggs slightly, covering the eggs with a halocarbon oil, and injecting the posterior end of the egg near the primordial germ cells (pole cells) with a 14- to 16-μm glass needle.[46,47] Injected embryos are then placed in a humid environment until hatching occurs. In several cases, removal of the egg chorion was not possible; however, some survival was obtained by injection through an intact chorion. Studies of embryo injection methodologies in nondrosopholphilid insects has been mostly limited to the analysis of transient gene expression due to the absence of functional gene vectors for these organisms. Table 3 lists various insect species successfully transfected via embryo injection.

As would be expected, insect embryo injection is a time-consuming, exacting procedure yielding varying success rates. Under optimal conditions, about 80% of injected *Drosophila* embryos hatch, and 40 to 50% survive to adulthood. Anywhere from 10 to 90% of the surviving adults may have injection-induced abnormalities or be sterile because of incompatibilities with their genetic backgrounds.[48] The survival to adulthood of injected embryos of other insects is in the same range as that obtained with *Drosophila*. For example, 6 to 15% of three mosquito species and 3.2% of a silkworm moth species developed to adults after an injection during the embryonic stage.[49-52]

A related microinjection procedure, injection of the egg within a gravid female mite, *Metaseiulius occidentalis*, has provided a means of obtaining stable transformation of this organism.[45] A 0.5% transformation frequency was obtained without using a vector system, and survival of the injected progeny was reported to be an order of magnitude higher than that of injected mosquito embryos. Whether the stable transformation gained by this method is related to the unique reproductive biology of this phytoseiid mite or to the introduction of exogenous DNA before fertilization of the egg is not yet known. However, further study is warranted, since this method may have application for transforming certain oviparous insects such as the tsetse fly.

The physical manipulations of the embryos necessary to accomplish the injection method have been implicated as influencing expression of heterologous genes linked to the *Drosophila* hsp70 promoter. Eberlein and Mitchell[53] suggest that stresses such as the dechorionation, desiccation, and injection steps induce subsequent expression of hspCAT activity in *Drosophila* embryos above that which would be expected to be expressed constitutively in the absence of a heat shock.

B. ALTERNATIVE DNA DELIVERY METHODS

The development of methods which would allow for the delivery of gene constructs into large numbers of embryos would permit the use of less efficient gene-transfer systems such as those typically employed in cell culture transformation. Successful transfection of *Drosophila* and *Anopheles gambiae* embryos using a bombardment technique have recently been reported. Balderelli and Lengyel[54] coated particles 1.2 µm in diameter with a DNA construct and blasted the microprojectiles shotgun-like into samples of 10 to 20,000 dechorionated, preblastoderm *Drosophila* embryos using a commercially available bioparticle delivery system. Mialhe, Laughinghouse, and Miller[55] used a similar procedure to treat samples containing 3 to 5000 mosquito embryos. In the latter study, the best results were gained without removing the egg chorion, but before hardening had occurred (60 to 80 min postoviposition). Survival to adulthood was not reported in either study, and the frequency of transfection was variable but repeatable.

Electroporation is another method which also has the potential to simultaneously deliver a gene construct to large numbers of embryos. Electroporation is a procedure where the insect embryos are suspended between two electrodes in a liquid medium containing the DNA construct and then are given a single or a series of high-voltage pulses. Transient holes or pores are produced in the egg shell and/or embryonic membranes through which the DNA is able to enter. Kamdar, Von Allmen, and Finnerty[56] obtained a 95% frequency of transient expression of the aldehyde oxidase gene transfected into *ma-1* mutant *Drosophila* embryos. They obtained these results without removal of the egg chorion and reported that about 75% of the treated embryos survived to third instar larvae. Dechorionation reduced survival to 10%, but the frequency of expression rose to 100%.

The effectiveness of the biolistic and electroporation methods apparently involves several interrelated factors: the chorion thickness and hardness, the embryonic stage of development, and the ability to survive treatment after chorion removal and/or penetration. It remains to be seen whether these two methods can be modified to accommodate insects where removal of a relatively thick egg shell is not possible or when dechorination exposes an extremely fragile egg membrane system.

Viral infection is a method of DNA delivery that has not been exploited in insects. This might be an especially appropriate method for hymenopteran parasitoids. Many of these parasitoid wasps transmit viruses of the polydnavirus family to their hosts.[57] These viruses appear to affect the immunologic mechanisms of the host in such a way as to allow parasitization by the wasp. What makes them especially interesting, however, is the fact that the viral DNA sequences appear to integrate into the genomic DNA of the parasitoid.[58] If the mechanisms of how this is accomplished can be determined, these types of viruses might be useful as natural transformation vectors.

C. ENHANCEMENT OF DNA DELIVERY

When stable transformation is the desired goal, strategies having the potential to increase the probability of gene integration into germline cells will enhance the success of any transfection scheme. Injection of gene constructs conjugated with yolk proteins into the hemocoel of female insects during egg development might be a means of obtaining selective uptake by oocytes prior to their fertilization.[59] Other strategies consist of transfection of the sperm within the male insect by injection of the developing testes[45] and *in vitro* transfection of sperm before artificial insemination (i.e., honey bee) or before employing *in vitro* fertilization.[60] Furthermore, a cationic liposome DNA carrier was found to be a successful means to transfect murine sperm,[61] but this method did not yield transgenic mice when the transfected sperm were tested via *in vitro* fertilization. In insects, foreign DNA was shown to bind externally to sperm of *Lucilla cuprina* and *Apis mellifera* without the benefit of a chemical mediator.[62]

The transplantation of primordial germ cells (pole cells) between mutant lines of *Drosophila* has been used to create germline mosaics for genetic and developmental biology studies. Increasing the production of transgenic insects may be possible by the *in vitro* transfection of pole cells which have been collected by the mass isolation technique of Allis, Waring, and Mahowald.[63] Complexing the gene construct with a chemical mediator for DNA, such as cationic liposomes, polybrene, DEAE-dextran, or polylysine, before injection, electroporation, or bombardment of the embryos could increase the efficiency of transfection. The chemical mediator would hopefully extend the life of an introduced DNA construct through several cycles of cleavage divisions of the preblastoderm embryo and thus increase the chances for integration when using a gene vector or by random integration.

VII. TRANSPOSABLE ELEMENT VECTORS

Transposable element vectors so far have proven to be the most efficient method for genetic transformation. Since P elements appear to be very species specific,[64] it may prove to be necessary to use an endogenous transposon system for each species. To develop such a system for insect parasitoids the first step must be to find a usable and functional transposable element from the species of choice. Table 4 lists seven different ways that have been used to identify and isolate transposable sequences.

TABLE 4
Methods to Find Transposable Elements

1. Sequence homology with known elements
2. Insertion caused mutations
3. Hybrid dysgenesis-like syndrome
4. Baculovirus insertion mutations
5. Foldback DNA
6. Sequence homology with reverse transcriptase
7. Middle repetitive DNA

The simplest method is simply looking for sequence homology with known elements using hybridization or PCR. This has been successful with the mariner family of elements,[65] but less so with others. It also has the disadvantage of detecting nonfunctional as well as functional elements. Many of the first transposable elements were identified by phenotypic mutations caused by their insertion into specific genes of *D. melanogaster* and *Zea mays*.[66,67] The *Drosophila* P elements were isolated because of a syndrome called hybrid dysgenesis caused by the sudden movement of the elements within the genome when a P-element containing strain mated with a strain not having P elements.[68] In lepidopteran insects, several transposable elements have been isolated from infecting baculoviruses.[69] The moving DNA elements insert into the viral genome and cause mutations in the viral phenotype. Some transposable elements have DNA sequences which are inverted repeats and which can form hairpin loops upon renaturation of denatured DNA. Potter[70] used this renaturation to isolate the foldback family of elements from *Drosophila*. Homology with reverse transcriptase of retroviruses has been used to isolate transposable elements which transpose via an RNA intermediate. Finally, Wichman, Pine, and Potter[71] used cross-hybridization of DNA from different species to find unique middle repetitive DNA sequences with the properties of transposable elements.

The primary advantage in the use of transposons lies in the fact that DNA introduced via a transposon vector system can be randomly and stably inserted into the recipient genome without necessarily adversely affecting the expression of other genes. Furthermore, transposable elements spread rapidly throughout an insect population once introduced, even in those cases whereby the fitness of the individual harboring the transposon is significantly reduced.[72] For example, if an element has a transposition efficiency of 80%, it could be lethal to 30% of its hosts and still spread to the entire population.[72] This is strictly due to the fact that when a transposable element-bearing individual mates with a wild-type counterpart, 100% of the F_1 generation has the potential of carrying the introduced DNA (as a result of transposition to multiple genomic sites), resulting in all of the chromosomes carrying at least one copy of the desired DNA.

Transposable elements are structurally diverse and can be divided into several groups based on their mode of transposition, similarities in their DNA sequence, and other molecular biological properties.[73-75] The four basic groups are (1) LTR-retrotransposons; (2) poly(A) retrotransposons; (3) foldback elements; and (4) short inverted repeat elements. From a biotechnological standpoint, a given member from

TABLE 5
Insect-Derived LTR-Retrotransposons

Transposon	Source	Size (kb)	Ref.
B104/roo	*Drosophila melanogaster*	8.7	76
Bari-1	*D. melanogaster*	1.7	100
Beagl	*D. melanogaster*	7.3	101
BEL	*D. melanogaster*	7.3	102
Blood	*D. melanogaster*	7.4	74
Copia	*D. melanogaster*	5.15	88
DhMiF2	*D. hydei*	4.8	103
DhMiF8	*D. hydei*	4.4	103
DM-1	*D. melanogaster*	5.3	103
DM-2	*D. melanogaster*	5.5	78
Doc	*D. melanogaster*	4.7	104
Flea	*D. melanogaster*	6.2	75
Gypsy	*D. melanogaster*	7.3	105
Kermit	*D. melanogaster*	4.8	106
mdg-1	*D. melanogaster*	7.3	107
mdg-3	*D. melanogaster*	5.4	108
mag	*Bombyx mori*	4.5	109
Minos	*D. hydei*	1.8	110
NEB	*D. melanogaster*	5.5	111
opus	*D. melanogaster*	4.8	74
Pao	*D. melanogaster*	5.1	74
Springer	*D. melanogaster*	8.8	112
Stalker	*D. melanogaster*	7.2	113
TECth1	*Chironomus thummi*	1.7	114
Tram	*D. miranda*	2.5	75
Trim	*D. miranda*	3.1	75
TED	*Trichoplusia ni*	17.6	115
Ulysses	*D. virilis*	10.7	116
17.6	*D. melanogaster*	7.4	101
297	*D. melanogaster*	7.0	117
412	*D. melanogaster*	7.0	118
1731	*D. melanogaster*	4.4	106
3S18	*D. melanogaster*	6.5	119

each family may prove to be useful for introducing desired DNA. Therefore, in an attempt to better understand the various types of transposons, a brief description of each family follows, as do tables listing the known members of each family of insect-derived transposable elements. For the sake of consistency, only those members for which a complete DNA sequence has been determined have been included.

A. LTR-RETROTRANSPOSONS

The majority of transposable elements identified in *D. melanogaster* transpose by means of an RNA intermediate by a mechanism similar to that utilized by retroviruses and are therefore referred to as retrotransposons (Table 5).[75,76] The retrotransposon lifecycle, as does the retrovirus lifecycle, involves both element-encoded functions and host cell mechanisms. Unlike retroviruses, however, retrotransposons have not

TABLE 6
Insect-Derived Poly(A)-Retrotransposons

Transposon	Source	Size	Ref.
Calypso	*Drosophila melanogaster*	7.2 kb	104
Delta88	*D. melanogaster*	7.0 kb	119
D	*D. melanogaster*	0.38 kb	120
Dong	*Bombyx mori*	3.0 kb	121
F	*D. melanogaster*	4.7 kb	77
G	*D. melanogaster*	4.0 kb	122
Harvey	*D. melanogaster*	7.2 kb	104
He-TA	*D. melanogaster*	9.4 kb	123
I	*D. melanogaster*	5.4 kb	124
IFP2	*Trichoplusia ni*	2.5 kb	125
ISY3	*D. miranda*	713.0 bp	75
Jockey	*D. melanogaster*	5.0 kb	79
LOA	*D. silvestris*	7.7 kb	126
NLR1Cth	*Chironomus thummi*	4.8 kb	127
R1AM	*Apis mellifera*	1.5 kb	78
1BM	*B. mori*	1.6 kb	78
R1DM	*D. melanogaster*	1.5 kb	128
R1NvA	*Nasonia vitripennis*	1.6 kb	129
R1NvB	*N. vitripennis*	1.6 kb	129
R1NvC	*N. vitripennis*	1.6 kb	129
R1NvD	*N. vitripennis*	1.5 kb	129
R1Pj	*Popilla japonica*	1.5 kb	129
R1Sc	*Sciara coprophila*	1.6 kb	129
R2BM	*B. mori*	2.1 kb	78
R2DM	*D. melanogaster*	2.1 kb	128
R2Pj	*P. japonica*	2.1 kb	129
R2Sc	*S. coprophila*	2.1 kb	129
R2Nv	*N. vitripennis*	1.8 kb	129
T1	*Anopheles gambiae*	4.6 kb	80
Wallaby	*D. melanogaster*	5.0 kb	104

been shown to be able to transgress cell boundaries, most likely due to their incapability of encoding capsid and envelope proteins, as do retroviruses.

B. POLY(A) RETROTRANSPOSONS

Poly(A) retrotransposons have been referred to as either the non-LTR class or the more appropriate poly(A)-retrotransposon family (Table 6). These elements are generally characterized by the absence of a terminal repeat, the presence of a polyadenylate sequence of varying lengths located at the 3′ end of the element, and variable sizes of target site duplication, as well as the presence of multiple truncated copies of the element at their 5′ end, suggesting that the reverse transcriptase processed minus strand synthesis is frequently abortive.[77] These elements, like their LTR-counterparts, are fairly widespread in nature. For instance, with the R1 and the R2 elements (designated BmR1, BmR2, DmR1, etc.), although they have only been identified in two species (i.e., cloned and sequenced), Southern blot hybridization

TABLE 7
Insect-Derived Foldback Transposable Elements

Transposon	Source	Size (kb)	Ref.
BS	*Drosophila melanogaster*	8	134
FB-4	*D. melanogaster*	10.0	70
FB-HB1	*D. melanogaster*	14.0	74
TE146	*D. melanogaster*	14.6	81
TFB1	*Chironomus thummi*	0.7	82

reveals their presence in over 40 different insect species from the grasshopper (*Melanoplus fermurrubrum*) to the parasitic wasp (*Trichomolopsis dubius*). There appears to be two different categories of poly(A)-type elements: those showing insertion site specificity and those appearing to insert randomly (DmR1 and Jockey are respective members of each category).[78,79]

C. FOLDBACK ELEMENTS

Foldback elements are transposable elements which are characterized by a uniquely complex structure (Table 7). Members of this small family of elements are characterized by long, modularly constructed, terminal inverted repeats flanking a variable central region. The inverted repeats themselves consist of variable numbers of tandemly repeated but imperfect copies of a 31-bp sequence. These repeated regions can vary significantly in size from several hundred basepairs to several kilobases. Additionally, the central portion of a foldback element can be quite diverse. Some elements have been identified which carry no additional insert DNA in several kilobases.[70,74,80-82]

D. SHORT INVERTED REPEAT ELEMENTS (SIRES)

Transposable elements of this classification (Table 8) have several features in common. These include short, inverted, terminal repeats with very weak sequence similarity and short (6 to 12 bp) genomic DNA duplications adjacent to the terminal repeats. Unlike retrotransposons, these elements lack long, terminal repeat sequences, as well as open reading frames corresponding to the various retrotransposon-encoded proteins. In general, a genome will either harbor a few copies of the autonomous (transposase-competent) full-length element and numerous copies of internally deleted defective elements or will totally lack the element. Although the number of members of this group appears to be quite small, three of the most "popular" *Drosophila* transposons belong to this group: mariner, hobo, and P element.[83-86]

E. FUTURE PROSPECTS FOR TRANSPOSONS

To produce a transgenic insect, it is imperative to introduce the foreign DNA in such a way as to achieve maximum integration frequency. To date, transformation systems based on P, hobo, and mariner elements have been developed for *Drosophila*, and P elements carrying the *Escherichia coli lacZ* gene have been used to detect enhancers stimulating gene expression in specific cell types.[14,87] The func-

TABLE 8
Insect-Derived Transposons with Short Inverted Repeat Termini

Transposon	Source	Size (kb)	Ref.
Hobo	*Drosophila melanogaster*	3	83
	D. simulans	2.6	84
Mariner	*D. mauritiana*	1.3	85
	D. simulans	1.3	85
	D. sechelli	1.3	85
	D. yakuba	1.3	85
	D. teissiera	1.3	85
MLE	*Chironomus plorabunda*	1.3	86
	Hyalaphord cecropia	1.25	130
P-element	*Chironomus procnemis*	3.0	89
	D. melanogaster	3.0	90
	D. willistoni	2.9	90
	Scaptomyza pallida	2.9	131
Pogo	*D. melanogaster*	2.1	132
Sancho-2	*D. melanogaster*	2.6	79
Uhu	*D. melanogaster*	2.5	133

tional properties of element transformation appear to be identical for these three transposons. From a biotechnological, as well as from a practical approach, members of the SIRE family represent the most likely candidates for successful genome manipulation. Although the P element has been shown to be quite species specific, there is reason to believe that either mariner or hobo, or other such related elements, will prove to be universally adaptable as a vector system for enacting specific desired changes.[86,88,89]

VIII. SELECTION OF TRANSFORMANTS

Once the exogenous DNA has been delivered into the target insect, it is necessary to be able to select those insects which have taken up the DNA construct into their genomic DNA. This selection must be dominant, completely penetrant, and easily scorable. With *Drosophila*, the method of choice has been the rescue of visible phenotypic mutations by coinjection of a selection gene along with the target gene. Eye color, rosy[15] or white,[90] has been the most used, while body color vectors have also been developed.[91] Insects where such genetically well-characterized phenotypes are not available will require other types of selection.

Drug resistance selection has been the most often-tried method. Resistance to the antibiotic neomycin can be conferred by expression of a bacterial neomycin phosphotransferase (NPT) gene that is not found in higher organisms. Coinjection of an NPT gene should confer dominant resistance to the drug which can be used to eliminate the nontransformed insects. Vectors have been designed with this in mind.[92] Neomycin resistance, however, has not proven to work as well in practice as in theory. It can have variable effects, and there seems to be a low level of natural resistance which can be unwittingly selected. Other systems involving the drugs hygromycin[93] and methotrexate[94] are also being investigated.

Metabolic enzyme selection systems such as alcohol dehydrogenase are also possible.[95] These, however, require either isolation or construction of null mutations for the enzyme, which could be difficult in genetically poorly characterized species.

IX. SUMMARY

We have discussed a number of possible ways in which biotechnology could be used in developing new and innovative insect management programs, particularly using insect parasitoids. While these technologies are realizable, they will require a considerable amount of new research. The base of knowledge surrounding the genetics, biochemistry, and developmental biology of not only parasitoids, but also their hosts needs to be significantly increased.

The development of a genetic transformation method is the keystone to further advances. Current transformation is limited to *Drosophila*. To realize its potential, transformation must be extended to the economically and medically important insects. We have attempted to examine possible directions to take toward attaining this goal and have pointed out potential pitfalls.

Genetic transformation of economically important insects will require the integration of several different lines of work (Figure 1). First, a solid base of knowledge of the genetics and cellular biology of the parasitoid Hymenoptera must be obtained. These genetic studies must be both classical and molecular. Development of mutant strains for selective purposes will be necessary, as will a thorough knowledge of the potential genetic diversity in wild populations. Second, the proper genes to achieve the desired trait must be isolated and modified. Third, a method to deliver the modified DNA to the nuclei of germline cells must be found. This involves not only the physical introduction of the DNA but its packaging in a vector which will insure its insertion into the chromosome and proper expression. Fourth, there must be a selection system so that the investigator can follow expression of the gene in the insects that have been transformed. Finally, before any release into the environment, the modified insects must be thoroughly studied for stability, effectiveness, and safety.

REFERENCES

1. **Crampton, J. M. and Eggleston, P., Eds.,** *Insect Molecular Science*, Academic Press, London, 1992.
2. **Whitten, M. J.,** The conceptual basis for genetic control, in *Comprehensive Insect Physiology, Biochemistry and Pharmacology*, Vol. 10, Kerkut, G. A. and Gilbert, L. I., Eds., Pergamon Press, Oxford, 1985, chap. 14.
3. **Crampton, J. M.,** Potential application of molecular biology in entomology, in *Insect Molecular Science*, Crampton, J. M. and Eggleston, P., Eds., Academic Press, London, 1992, chap. 1.

4. **Ashburner, M.,** Mapping insect genomes, in *Insect Molecular Science*, Crampton, J. M. and Eggleston, P., Eds., Academic Press, London, 1992, chap. 4.
5. **Botstein, D., White, R. L., Skolnick, M., and Davis, R. W.,** Construction of a genetic linkage map in man using restriction fragment length polymorphisms, *Am. J. Hum. Genet.*, 32, 314, 1980.
6. **Heckel, D. G.,** Comparative genetic linkage mapping in insects, *Annu. Rev. Entomol.*, 38, 381, 1993.
7. **Love, J. M., Knight, A. M., McAleer, M. A., and Todd, J. A.,** Towards construction of a high resolution map of the mouse genome using PCR-analysed microsatellites, *Nucl. Acids Res.*, 18, 4123, 1990.
8. **Roehrdanz, R. L., Reed, D. K., and Burton, R. L.,** Use of polymerase chain reaction and arbitrary primers to distinguish laboratory-raised colonies of parasitic hymenoptera, *Biol. Control*, 3, 199, 1993.
9. **Dushay, M. S., Rosbash, M., and Hall, J. C.,** Mapping the clock rhythm mutation to the period locus of *Drosophila melanogaster* by germline transformation, *J. Neurogenet.*, 8, 173, 1992.
10. **Davis, R. L. and Dauwalder, B.,** The *Drosophila* dunce locus: learning and memory genes in the fly, *Trends Genet.*, 7, 224, 1991.
11. **Feyereisen, R., Koener, J. F., Farnsworth, D. E., and Nebert, D. W.,** Isolation and sequence of cDNA encoding a cytochrome P-450 from an insecticide-resistant strain of the house fly, *Musca domestica*, *Proc. Natl. Acad. Sci. U.S.A.*, 86, 1465, 1989.
12. **ffrench-Constant, R. H., Steichen, J. C., Rocheleau, T. A., Aronstein, K., and Roush, R. T.,** A single amino acid substitution in a aminobutyric acid subtype A receptor locus is associated with cyclodiene insecicide resistance in *Drosophila* populations, *Proc. Natl. Acad. Sci. U.S.A.*, 90, 1957, 1993.
13. **Salama, M. S. and Miller, T. A.,** *In vitro* translation of diapause mRNA from the fat body of active and diapause larvae of the pink bollworm, *Pectinophora gossypiella* Saunders, *Arch. Insect Biochem. Physiol.*, 23, 1, 1993.
14. **Sato, Y., Oguchi, M., Menjo, N., Imai, K., Saito, H., Ikeda, M., Isobe, M., and Yamashita, O.,** Precursor polyprotein for multiple neuropeptides secreted from the subesophageal ganglion of the silkworm *Bombyx mori*: characterization of the cDNA encoding the diapause hormone precursor and identification of additional peptides, *Proc. Natl. Acad. Sci. U.S.A.*, 90, 3251, 1993.
15. **Spradling, A. C. and Rubin, G. M.,** Transposition of cloned P elements into *Drosophila* germ line chromosomes, *Science*, 218, 348, 1982.
16. **Graur, D.,** Gene diversity in Hymenoptera, *Evolution*, 39, 190, 1985.
17. **DeBach, P. and Rosen, D., Eds.,** *Biological Control by Natural Enemies*, Cambridge University Press, Cambridge, U.K., 1991.
18. **Hoy, M. A., Westigard, P. H., and Hoyt, S. C.,** Release and evaluation of a laboratory selected strain of the predaceous mite *Metaseiulus occidentalis* (Acari: Phytoseiidae) in southern Oregon pear orchards and a Washington apple orchard, *J. Econ. Entomol.*, 76, 383, 1983.
19. **Mackey, T. F.,** Insertional mutagenesis and quantitative variation in *Drosophila*, *Volume of Abstracts: Seventeenth International Congress of Genetics*, Genetic Society of America, 1993, 1.
20. **Nebert, D. W. and Gonzalez, F. J.,** P450 genes: structure, evolution and regulation, *Annu. Rev. Biochem.*, 56, 945, 1986.
21. **Lagadic, L., Cuany, A., Berge, J.-B., and Echaubard, M.,** Purification and partial characterization of glutathione S-transferases from insecticide resistant and lindane-induced susceptible *Spodoptera littoralis* (Boisd.) larvae, *Insect Biochem. Mol. Biol.*, 23, 467, 1993.
22. **Raymond, M., Poulin, E., Boiroux, V., Dupont, E., and Pasteur, N.,** Stability of insecticide resistance due to amplification of esterase genes in *Culex pipiens*, *Heredity*, 70, 301, 1993.
23. **Taylor, M. F. J., Heckel, D. G., Brown, T. M., Kreitman, M. E., and Black, B.,** Linkage of pyrethroid insecticide resistance to the sodium channel locus in the tobacco budworm, *Insect Biochem. Mol. Biol.*, 23, 763, 1993.
24. **Oppenoorth, F. J.,** Biochemistry and genetics of insecticide resistance, in *Comprehensive Insect Physiology, Biochemistry and Pharmacology*, Vol. 12, Kerkut, G. A. and Gilbert, L. I., Eds., Pergamon Press, Oxford, 1985, chap. 19.

25. **Rosenheim, J. A. and Hoy, M. A.**, Genetic improvement of a parasitoid biological control agent: artificial selection for insecticide resistance in *Aphytis melinus* (Hymenoptera: Aphelinidae), *J. Econ. Emtomol.*, 81, 1539, 1988.
26. **Hoy, M. A. and Cave, F. E.**, Glutathione-resistant strain of walnut aphid parasite, *Calif. Agric.*, 42, 4, 1988.
27. **Schmidt, O. and Theopold, U.**, Immune defense and suppression in insects, *Bioessays*, 13, 343, 1991.
28. **Kimbrell, D. A.**, Insect antibacterial proteins: not just for insects and against bacteria, *Bioessays*, 13, 657, 1991.
29. **Samakovlis, C., Kimbrell, D. A., Kylsten, P., Engstrom, A., and Hultmark, D.**, The immune response in *Drosophila*: pattern of cecropin expression and biological activity, *EMBO J.*, 9, 2969, 1990.
30. **Denlinger, D. L.**, Relationship between cold hardiness and diapause, in *Insects at Low Temperature*, Lee, R. E. and Denlinger, D. L., Eds., Chapman and Hall, New York, 1991, 174.
31. **Duman, J. G., Horwarth, K. L., Tomchaney, A., and Patterson, J. L.**, Antifreeze agents of terrestrial arthropods, *Comp. Biochem. Physiol. A*, 73, 545, 1982.
32. **Hoy, M. A.**, Rapid response to selection for a nondiapausing gypsy moth, *Science*, 196, 1462, 1977.
33. **Salama, M. S. and Miller, T. A.**, A diapause associated protein of the pink bollworm *Pectinophora gossypiella* Saunders, *Arch. Insect Biochem. Physiol.*, 21, 1, 1992.
34. **Eguchi, Y., Itoh, T., and Tomizawa, J.**, Antisense RNA, *Annu. Rev. Biochem.*, 60, 631, 1990.
35. **Wrensch, D. L. and Ebbert, M. A.**, *Evolution and Diversity of Sex Ratio in Insects and Mites*, Chapman and Hall, New York, 1993.
36. **Flanders, S. E.**, The mechanism of sex ratio regulation in the parasitic Hymenoptera, *Insect Soc.*, 3, 325, 1956.
37. **King, B. H.**, Offspring sex ratios in parasitoid wasps, *Q. Rev. Biol.*, 62, 367, 1987.
38. **King, B. H.**, Sex-ratio manipulation in response to host size by the parasitoid wasp *Spalangia cameroni*: a laboratory study, *Evolution*, 42, 1190, 1988.
39. **Nur, U., Werren, J. H., Eickbush, D., and Burke, W. D.**, A selfish B chromosome that enhaces its transmisson by eliminating the paternal genome, *Science*, 240, 512, 1988.
40. **Steinman-Zwicky, M., Amrein, H., and Nothiger, R.**, Genetic control of sex determination in *Drosophila*, *Adv. Genet.*, 27, 189, 1990.
41. **Ebbert, M. A.**, Endosymbiotic sex ratio distorters in insects and mites, in *Evolution and Diversity of Sex Ratio in Insects and Mites*, Wrensch, D. L. and Ebbert, M. A., Eds., Chapman and Hall, New York, 1993, chap. 4.
42. **Werren, J. H., Skinner, S. W., and Charnov, E. L.**, Paternal inheritance of a daughterless sex ratio factor, *Nature (London)*, 293, 467, 1981.
43. **Skinner, S. W.**, Maternally inherited sex ratio in the parasitoid wasp, *Nasonia vitripennis*, *Science*, 215, 1133, 1982.
44. **Skinner, S. W.**, Son-killer: a third extrachromosomal factor affecting the sex ratio in the parasitoid wasp, *Nasonia vitripennis*, *Genetics*, 109, 745, 1985.
45. **Presnail, J. K. and Hoy, M. A.**, Stable genetic transformation of a beneficial arthropod, *Metaseiulus occidentalis* (Acari: Phytoseiidae), by a microinjection technique, *Proc. Natl. Acad. Sci. U.S.A.*, 89, 7732, 1992.
46. **van Deusen, E. B.**, Sex determination in germ line chimeras of *Drosophila melanogaster*, *J. Embryol. Exp. Morphol.*, 37, 173, 1976.
47. **Ashburner, M.**, *Drosophila — A Laboratory Handbook*, Cold Spring Harbor Laboratory Press, Cold Spring Harbor, New York, 1989, 228.
48. **Pirrotta, V.**, Vectors for P-mediated transformation in *Drosophila*, in *Vectors — A Survey of Molecular Cloning Vectors and Their Uses*, Rodriguez, R. L. and Denhardt, D. T., Eds., Butterworths, Boston, 1988, 437.
49. **Miller, L. K., Sakai, R. K., Romans, P., Gwadz, R. W., Kautoff, P., and Coon, H. G.**, Stable integration of a bacterial gene in the mosquito, *Anopheles gambiae*, *Science*, 237, 779, 1987.

50. **McGrane, V., Carlson, J. O., Miller, B. R., and Beaty, B. J.,** Microinjection of DNA into *Aedes triseriatus* ova and detection of integration, *Am. J. Trop. Med. Hyg.*, 39, 502, 1988.
51. **Morris, A. C., Eggleston, P., and Crampton, J. M.,** Genetic transformation of the mosquito, *Aedes aegypti*, by microinjection of DNA, *Med. Vet. Entomol.*, 3, 1, 1989.
52. **Nikolaev, A. I., Chkoniya, T. I., and Kafiani-Eristavi, K. A.,** Extrachromosomal localization and hereditary transfer of recombinant plasmid microinjected into silkworm eggs, *Mol. Biol.*, 25, 896, 1991.
53. **Eberlein, S. and Mitchell, H. K.,** Protein synthesis patterns following stage-specific heat shock in early *Drosophila* embryos, *Mol. Gen. Genet.*, 210, 407, 1987.
54. **Baldarelli, R. M. and Lengyel, J. A.,** Transient expression of DNA after ballistic introduction into *Drosophila* embryos, *Nucl. Acids Res.*, 18, 5903, 1992.
55. **Mialhe, E., Laughinghouse, A., and Miller, L. H.,** Biolistics for gene transfer to mosquito embryos, *Proc. FAO/IAEA Int. Symp. Management Insect Pests: Nuclear Rel. Mol. Genet. Tech.*, International Atomic Energy Agency, Vienna, 1993.
56. **Kamdar, P., Von Allmen, G., and Finnerty, V.,** Transient expression of DNA in *Drosophila* via electroporation, *Nucl. Acids Res.*, 20, 3526, 1993.
57. **Fleming, J. G. W.,** Polydnaviruses: mutualists and pathogens, *Annu. Rev. Entomol.*, 37, 401, 1992.
58. **Fleming, J. G. W. and Summers, M.D.,** Polydnavirus DNA is integrated in the DNA of its parasitoid wasp host, *Proc. Natl. Acad. Sci. U.S.A.*, 88, 9770, 1991.
59. **Handler, A. M. and O'Brochta, D. A.,** Prospects for gene transformation in insects, *Annu. Rev. Entomol.*, 36, 159, 1991.
60. **Sawa, M.,** Fertilization of hetero-specific insect eggs by sperm injection, *Jpn. J. Genet.*, 66, 297, 1991.
61. **Bachiller, D., Schellander, K., Peli, J., and Ruther, U.,** Liposome-mediated DNA uptake by sperm cells, *Mol. Reprod. Dev.*, 30, 194, 1991.
62. **Atkinson, P. W., Hines, E. R., Beaton, S., Matthaei, K. I., and Reed, K. C.,** Association of exogenous DNA with cattle and insect spermatozoa *in vitro*, *Mol. Reprod. Dev.*, 29, 1, 1991.
63. **Allis, C. D., Waring, G. L., and Mahowald, A. P.,** Mass isolation of pole cells from *Drosophila melanogaster*, *Dev. Biol.*, 56, 372, 1977.
64. **O'Brochta, D. A. and Handler, A. M.,** Mobility of P elements in *drosophilids* and non-*drosophilids*, *Proc. Natl. Acad. Sci. U.S.A.*, 85, 779, 1988.
65. **Robertson, H. M.,** The mariner transposable element is widespread in insects, *Nature (London)*, 362, 241, 1993.
66. **Bingham, P. M., Levis, R., and Rubin, G. M.,** The cloning of the DNA sequences from the white locus of *Drosophila melanogaster* using a novel and general method, *Cell*, 25, 693, 1981.
67. **Fedoroff, N.,** About *Maize* transposable elements and development, *Cell*, 56, 181, 1989.
68. **Bregliano, J. C. and Kidwell, M. G.,** Hybrid dysgenesis determinants, in *Mobile Genetic Elements*, Shapiro, J., Ed., Academic Press, New York, 1983, 363.
69. **Friesen, P. D. and Nissen, M. S.,** Gene organization and transcription of TED, a lepidopteran retrotransposon integrated within the baculovirus genome, *Mol. Cell. Biol.*, 10, 3067, 1990.
70. **Potter, S. S.,** DNA sequence of a foldback transposable element in *Drosophila*, *Nature (London)*, 297, 201, 1982.
71. **Wichman, H. A., Pine, D. S., and Potter, S. S.,** Mys, a family of mammalian transposable elements isolated by phylogenetic screening, *Nature (London)*, 353, 864, 1985.
72. **Kidwell, M. G.,** Evolutionary aspects of hybrid dysgenesis in *Drosophila*, *Can. J. Zool.*, 68, 1716, 1989.
73. **Gelbart, W. M. and Blackman, R. K.,** The hobo element of *Drosophila melanogaster*, *Prog. Nucl. Acid Res. Mol. Biol.*, 36, 37, 1989.
74. **Bingham, P. M. and Chapman, C. H.,** Evidence that white-blood is a novel type of temperature-sensitive mutation resulting from temperature-dependent effects of a transposon insertion on formation of white transcripts, *EMBO J.*, 5, 3343, 1986.
75. **Finnegan, D. J.,** Transposable elements, *Curr. Opin. Genet. Dev.*, 2, 861, 1992.
76. **Scherer, G., Tschudi, C., Perera, J., Delius, H., and Pirrotta, V.,** B104 a new dispersed repeated gene family in *Drosophila melanogaster* and its analogies with retroviruses, *J. Mol. Biol.*, 157, 435, 1982.

77. **Di Nocera, P. P., Digan, M. E., and Dawid, I.,** A family of oligo-adenylated transposable sequences in *Drosophila melanogaster*, *J. Mol. Biol.*, 168, 715, 1983.
78. **Xiong, Y. and Eickbush, T. H.,** Origin and evolution of retroelements based upon their reverse transcriptase sequences, *EMBO J.*, 9, 3353, 1990.
79. **Georgiev, G. P.,** Mobile genetic elements in animal cells and their biological significance, *Eur. J. Biochem.*, 145, 203, 1984.
80. **Besansky, N. J.,** A retrotransposable element from the mosquito *Anopheles gambiae*, *Mol. Cell. Biol.*, 10, 863, 1990.
81. **Lovering, R., Harden, N., and Ashburner, M.,** The molecular structure of TE146 and its derivatives in *Drosophila melanogaster*, *Genetics*, 128, 357, 1991.
82. **Hankeln, T. and Schmidt, E. R.,** New foldback transposable element TFB1 found in histone genes of the midge *Chironomus thummi*, *J. Mol. Evol.*, 215, 477, 1990.
83. **Mcginnis, W., Shermoen, A. W., and Beckendork, S. K.,** A transposable element inserted just 5′ to a *Drosophila* glue protein gene alters gene expression and chromatin structure, *Cell*, 34, 75, 1983.
84. **Boussy, I. A. and Daniels, S. B.,** Hobo transposable elements in *Drosophila melanogaster* and *D. simulans*, *Genet. Res. Camb.*, 58, 27, 1991.
85. **Maruyama, K. and Hartl, D. L.,** Evidence for interspecific transfer of the transposable element mariner between *Drosophila* and *Zaprionus*, *J. Mol. Evol.*, 33, 514, 1991.
86. **Robertson, H. M., Lampe, D. J., and Macleod, E. G.,** A mariner transposable element from a lacewing, *Nucl. Acids Res.*, 20, 6409, 1992.
87. **Rubin, G. M. and Spradling, A. C.,** Genetic transformation of *Drosophila* with transposable elements, *Science*, 218, 348, 1982.
88. **Finnegan, D. J., Rubin, G. M., Young, M. W., and Hogness, D. S.,** Repeated gene families in *Drosophila melanogaster*, *Cold Spring Harbor Symp. Quant. Biol.*, 42, 1053, 1978.
89. **Handler, A. M., Gomez, S. P., and O'Brochta, D. A.,** A functional analysis of the P-element gene-transfer vector in insects, *Arch. Insect Biochem. Physiol.*, 22, 373, 1993.
90. **O'Hare, K., Murphy, C., Levis, R., and Rubin, G. M.,** DNA sequence of the white locus of *Drosophila melanogaster*, *J. Mol. Biol.*, 180, 437, 1984.
91. **Patton, J. S., Gomes, X. V., and Geyer, P. K.,** Position-independent germline transformation in *Drosophila* using a cuticle pigmentation gene as a selectable marker, *Nucl. Acids Res.*, 20, 5859, 1992.
92. **Steller, H. and Pirrotta, V.,** A transposable P vector that confers selectable G418 resistance to *Drosophila* larvae, *EMBO J.*, 4, 167, 1985.
93. **van der Straten, A., Johansen, H., Sweet, R., and Rosenberg, M.,** Efficient expression of foreign genes in cultured *Drosophila melanogaster* cells using hygromycin B selection, in *Invertebrate and Fish Tissue Culture*, Kuroda, Y., Kurstak, E., and Maramorosch, K., Eds., Japan Scientific Societies Press, Tokyo, 1988, 131.
94. **Fallon, A. M.,** DNA mediated gene transfer: applications to mosquitoes, *Nature (London)*, 352, 828, 1991.
95. **Robinson, A. S.,** Sex-ratio manipulation in relation to insect pest control, *Annu. Rev. Genet.*, 17, 191, 1983.
96. **McInnis, D. O., Haymer, D. S., Tam, S. Y. T., and Thanaphum, S.,** *Ceratitis capitata* (Diptera: Tephritidae): transient expression of a heterologous gene for resistance to the antibiotic geneticin, *Ann. Entomol. Soc. Am.*, 83, 982, 1990.
97. **Atkinson, P. W. and O'Brochta, D. A.,** *In vivo* expression of two highly conserved *Drosophila* genes in the Australian sheep blowfly, *Lucillia cuprina*, *Insect Biochem. Mol. Biol.*, 5, 423, 1992.
98. **Tamura, T., Kanda, T., Takiya, S., Okano, K., and Maekawa, H.,** Transient expression of chimeric CAT genes injected into early embryos of the domesticated silkworm, *Bombyx mori*, *Jpn. J. Genet.*, 65, 401, 1990.
99. **Mathi, S., Walker, V. K., and Wyatt, G. R.,** Expression from two *Drosophila* promoters in embryos of the migratory locust, *Insect Biochem.*, 21, 623, 1991.
100. **Caizzi, R., Caggese, C., and Pimpinelli, S.,** Bar-1, a new transposon-like family in *Drosophila melanogaster* with a unique heterochromatic organization, *Genetics*, 133, 335, 1993.

101. **Snyder, M. P., Kimbrell, D., Hunkapiller, M., Hill, R., Fristrom, J., and Davidson, N.,** A transposable element that splits the promoter inactivates a *Drosophila* cuticle protein gene, *Proc. Natl. Acad. Sci. U.S.A.*, 79, 7430, 1982.
102. **Goldberg, M. L., Sheen, J., Gehring, W. J., and Green, M. M.,** Unequal crossing-over associated with asymmetrical synapsis between nomadic elements in the *Drosophila melanogaster* genome, *Proc. Natl. Acad. Sci. U.S.A.*, 79, 265, 1983.
103. **Lankenau, D. H. and Hennig, W.,** Micropia-Dm2, the nucleotide sequence of a rearranged retrotransposon from *Drosophila melanogaster*, *Nucl. Acids Res.*, 18, 4265, 1990.
104. **Driver, A., Lacey, S. F., Cullingfor, T. E., and O'Hare, K.,** Structural analysis of Doc transposable elements associated with mutations at the white and suppressor of forked loci of *Drosophila melanogaster*, *Mol. Gen. Genet.*, 220, 49, 1989.
105. **Freund, R. and Meselson, M.,** Long terminal repeat nucleotide sequence and specific insertion of the gypsy transposon, *Proc. Natl. Acad. Sci. U.S.A.*, 81, 4462, 1984.
106. **Peronnet, F., Becker, J. L., Becker, J., d'Auriol, L., Galibert, F., and Best-Belpomme, M.,** 1731, a new retrotransposon with hormone modulated expression, *Nucl. Acids Res.*, 22, 9017, 1986.
107. **Kulguskin, V. V., Ilyin, Y. V., and Georgiev, G. P.,** Mobile dispersed genetic element MDG1 of *Drosophila melanogaster*: nucleotide sequence of long terminal repeats, *Nucl. Acids Res.*, 9, 3451, 1981.
108. **Bayev, A. A., Krayev, A. S., Lyubomirskaya, N. V., Ilyin, Y. V., Skryabin, K. G., and Georgiev, G. P.,** The transposable element mdg3 in *Drosophila melanogaster* is flanked with perfect direct and mismatched inverted repeats, *Nucl. Acids Res.*, 8, 3263, 1980.
109. **Michaille, J.-J., Mathavan, S., Gaillard, J., and Garel, A.,** The complete sequence of mag, a new retrotransposon in *Bombyx mori*, *Nucl. Acids Res.*, 18, 674, 1990.
110. **Franz, G. and Savakis, C.,** Minos, a new transposable element from *Drosophila hydei*, is a member of the TC1-like family of transposons, *Nucl. Acids Res.*, 23, 6646, 1991.
111. **Paro, R., Goldberg, M. L., and Gehring, W.,** Molecular analysis of large transposable elements carrying the white locus of *Drosophila melanogaster*, *EMBO J.*, 2, 853, 1983.
112. **Karlic, C. C. and Fyrberg, E. A.,** An insertion within a variably spliced *Drosophila* tropomyosin gene blocks accumulation of only one encoded isoform, *Cell*, 41, 57, 1985.
113. **Georgiev, P. G., Kiselev, S. L., Simonova, O. B., and Gerasimova, T. I.,** A novel transposition system in *Drosophila melanogaster* depending on the Stalker mobile genetic element, *EMBO J.*, 9, 2037, 1990.
114. **Wobus, U., Baumlein, H., Bogachev, S. S., Borisevich, I. V., Panitz, R., and Kolesnikov, N.,** A new transposable element in *Chironomus thummi*, *Mol. Gen. Genet.*, 222, 311, 1990.
115. **Evgen'ev, M. B., Corces, V. G., and Lankenau, D. H.,** Ulysses transposable element of *Drosophila* shows high structural similarities to functional domains of retroviruses, *J. Mol. Biol.*, 225, 917 1992.
116. **Ikenaga, H. and Saigo, K.,** Insertion of mobile genetic element 297 into the T-A-T-A box for the H3 histone in *Drosophila melanogaster*, *Proc. Natl. Acad. Sci. U.S.A.*, 79, 4143, 1982.
117. **Will, B. M., Bayev, A. A., and Finnegan, D. J.,** Nucleotide sequence of terminal repeats of 412 transposable elements of *Drosophila melanogaster*, *J. Mol. Biol.*, 153, 897, 1981.
118. **Bell, J. R., Bogardus, A. M., Schmidt, T., and Pellegrini, M.,** A new copia-like transposable element found in a *Drosophila* rDNA gene unit, *Nucl. Acids Res.*, 13, 3861, 1985.
119. **Pittler, S. J. and Davis, R. L.,** A new family of the poly-deoxyadenylated class of *Drosophila* transposable elements identified by a representative member at the dunce locus, *Mol. Gen. Genet.*, 208, 325, 1987.
120. **Di Nocera, P. P. and Dawid, I. B.,** Interdigitated arrangement of two oligo(A)-terminated DNA sequences in *Drosophila*, *Nucl. Acids Res.*, 11, 5475, 1983.
121. **Xiong, Y. and Eickbush, T. H.,** Dong a non-long terminal repeat non-Ltr retrotransposable element from *Bombyx mori*, *Nucl. Acids Res.*, 21, 1318, 1993.
122. **Di Nocera, P. P. Graziani, F., and Lavorgna, G.,** Genomic and Structureal organization of Drosphila melanogaster G elements, *Nucl. Acids Res.*, 11, 5475, 1983.

123. **Biessmann, H., Valgeirsdottir, K., Lofsky, A., Chin, C., Ginther, B., Levis, R. W., and Pardue, M.-L.,** Het-A: a transposable element specifically involved in "healing" broken chromosome ends in *Drosophila melanogaster*, *Mol. Cell. Biol.*, 12, 3910, 1992.
124. **Pritchard, M. A., Dura, J. M., Pelisson, A., Bucheton, A., and Finnegan, D. J.,** A cloned I factor is fully functional in *Drosophila melanogaster*, *Mol. Gen. Genet.*, 214, 533, 1988.
125. **Cary, L. C., Goebel, M., Corsaro, B. G., Way, H.-G., Rosen, E., and Fraser, M. J.,** Transposon mutagenesis of baculoviruses: analysis of *Trichoplusia ni* transposon IFP2 insertions within the FP locus of nuclear polyhedrosis viruses, *Virology*, 172, 156–169, 1989.
126. **Felger, I. and Hunt, J. A.,** A non-LTR retrotransposon from the Hawaiian *Drosophila*: the LOA element, *Genetica*, 85, 119, 1992.
127. **Blinov, A. G., Sobanov, Y. V., Bogachev, S. S., Donchenko, A. P., and Filippova, M. A.,** The *Chironomus thummi* genome contains a non-LTR retrotransposon, *Mol. Gen. Genet.*, 237, 412, 1993.
128. **Jakubczak, J. L., Xiong, Y., and Eickbush, T. H.,** Type I(R1) and typeII (R2) ribosomal DNA insertions of *Drosophila melanogaster* are retrotransposable elements closely related to those of *Bombyx mori*, *J. Mol. Biol.*, 212, 37, 1990.
129. **Burke, W. D., Eickbush, D. G., Xiong, Y., Jakubczak, J., and Eickbush, T. H.,** Sequence relationship of retrotransposable elements R1 and R2 within and between divergent insect species, *Mol. Biol. Evol.*, 10, 163, 1993.
130. **Lidholm, D.-A., Gudmundsson, G. H., and Boman, H. G.,** A highly repetitive, mariner-like element in the genome of *Hyalophora cecropia*, *J. Biol. Chem.*, 266, 11518, 1991.
131. **Simonelig, M. and Anxolabehere, D.,** A P element of *Scaptomyza pallida* is active in *Drosophila melanogaster*, *Proc. Natl. Acad. Sci. U.S.A.*, 88, 6102, 1991.
132. **Tudor, M., Lobocka, M., Goodell, M., Pettitt, J., and O'Hare, K.,** The pogo transposable element family of *Drosophila melanogaster*, *Mol. Gen. Genet.*, 232, 126, 1992.
133. **Brezinsky, L., Wang, G. V. L., Humphreys, T., and Hunt, J.,** Transposable element Uhu from Hawaiian *Drosophila* is a member of the widely dispersed class of tc1-like transposons, *Nucl. Acids Res.*, 18, 2053, 1990.
134. **Campuzano, S., Balcells, L., Villares, R., Carramolino, L., Garcia-Alonso, L., and Modelell, J.,** Excess function Hairy-wing mutations caused by gypsy and copia transposable elements inserted within structural genes of the achaete-scute locus of Drosophila, *Cell*, 44, 303, 1986.
135. **Leopold, R. A.,** unpublished results, 1993.

INDEX

B

C

F

G

H

I

Q

R

U

V

Z